Why You
Shouldn't Eat
Your Boogers and
Other Useless or
Gross Information
About Your Body

Why You Shouldn't Eat Your Boogers and Other Useless or Gross Information About Your Body

FRANCESCA GOULD

Jeremy P. Tarcher/Penguin
a member of Penguin Group (USA) Inc.
New York

JEREMY P. TARCHER/PENGUIN
Published by the Penguin Group
Penguin Group (USA) Inc., 375 Hudson Street, New York,
New York 10014, USA • Penguin Group (Canada), 90 Eglinton Avenue East,
Suite 700, Toronto, Ontario M4P 2Y3, Canada (a division of Pearson
Canada Inc.) • Penguin Books Ltd, 80 Strand, London WC2R 0RL, England •
Penguin Ireland, 25 St Stephen's Green, Dublin 2, Ireland (a division of
Penguin Books Ltd) • Penguin Group (Australia), 250 Camberwell Road,
Camberwell, Victoria 3124, Australia (a division of Pearson Australia
Group Pty Ltd) • Penguin Books India Pvt Ltd, 11 Community Centre,
Panchsheel Park, New Delhi–110 017, India • Penguin Group (NZ), 67 Apollo Drive,
Rosedale, North Shore 0632, New Zealand (a division of Pearson
New Zealand Ltd) • Penguin Books (South Africa) (Pty) Ltd, 24 Sturdee Avenue,
Rosebank, Johannesburg 2196, South Africa

Penguin Books Ltd, Registered Offices:
80 Strand, London WC2R 0RL, England

Most Tarcher/Penguin books are available at special quantity discounts for bulk purchase
for sales promotions, premiums, fund-raising, and educational needs. Special books or
book excerpts also can be created to fit specific needs.
For details, write Penguin Group (USA) Inc. Special Markets,
375 Hudson Street, New York, NY 10014.

Library of Congress Control Number: 2008922332
ISBN 978-1-58542-645-4

Printed in the United States of America
3   5   7   9   10   8   6   4

Book design by Paul Saunders

This book is dedicated to my mum, Jayne, and my beautiful daughter, Charlotte

Contents

Acknowledgments

The author would like to thank Susan Punton for reviewing the text. Also David Haviland and Carl Cutler for their help in the production of this book.

CHAPTER ONE

Epidermal Ephemera

How can an ant's head be used to heal a wound?

Even today, some South American tribes use the mandibles of soldier ants to help heal cuts. The skin at the site of the wound is pushed together, and the ant is placed onto it so that its pincers dig into the flesh either side of the wound, like a stitch. Then, the ant's body is twisted off, so that only the head remains. If the wound is quite large, many ants may be used to help seal the cut, so that the end result resembles a row of stitches. As unusual as this may sound, this method has proved an effective way of helping wounds to heal. This method also used to be practiced in Africa and India.

How can maggots help to solve crime?

Insects, particularly blowflies and their maggots, can provide important evidence in the investigation of a murder. Adult blowflies have a great sense of smell, and they find the odor of decaying flesh extremely appealing. They colonize dead bodies rapidly after death, and because they are so quick, the size and age of blowfly maggots on a corpse can be used to measure the time, and sometimes the place, of death.

In Scotland in 1935, human remains were found dumped in a small ravine. Blowfly maggots were discovered on the decaying bodies. The remains were later identified as those of the wife and maid of Dr. Buck Ruxton of Lancaster. A doctor estimated the age of the maggots, and this provided a vital clue as to when the murders had taken place. Because of these maggots, Dr. Ruxton was found guilty of the murders and hanged.

The Body Farm in Tennessee is no ordinary farm. It is a center that studies the decomposition of dead bodies using real human corpses. They use bodies that have been donated to science and place them in a variety of situations such as in a car, under water, in a wooded area, buried, and so on, to see what bugs do to the corpses.

Why do we get goose bumps?

Goose bumps occur, when we are cold, because of the contraction of tiny muscles (erector pili muscles) in the skin that cause hairs to stand on end resulting in little bumps. In our hairier days lots of hairs would stand on end and would hold a layer of warm air between the skin and hair. The heat radiating from the skin would warm the trapped air and help to keep us warm.

Also, when our hair stood on end, it might have made us look bigger, or scarier. It is, perhaps, for this reason that we get goose bumps when we feel nervous or angry.

How does Botox work?

One way in which wrinkles are formed is through the contraction of our facial muscles, for example when we smile or squint. Along with other factors, repetitive use of these muscles can lead to the formation of lines on the face. Botox, which is injected into the skin, helps to reduce wrinkles, making the skin appear smoother. The effect of Botox is temporary, and another injection may be needed after just a few months.

Botox is derived from the botulinum toxin A, from the bacterium that produces botulism—an often fatal form of food poisoning. Botulism kills its victims by paralyzing the muscles normally used to breathe, causing suffocation. However, a Botox injection doesn't contain any of the botulism bacteria; rather, it contains a tiny amount of a chemical that is produced by the bacteria. Botox treatment appears to be very safe, and was originally used in the 1980s to treat eye disorders including crossed eyes and uncontrollable blinking.

Botox treatment consists of injecting Botox directly into a muscle, temporarily paralyzing it. Botox doesn't harm the muscle itself; instead, it works by affecting the nerves that control the muscle. The muscle becomes paralyzed because the Botox prevents the passing of the nerve messages that normally stimulate the muscle to contract. Botox can also be used to help prevent excessive sweating, and is used on children with cerebral palsy to prevent muscle tension and other symptoms.

What is a collagen injection?

Collagen is a natural substance that is found in our skin, muscles, tendons and bones. It provides structural support in the bodies of humans and animals.

There are various forms of collagen, some of which can be used to help smooth wrinkles or plump up our lips. Bovine collagen is extracted from the skin of dead cows, sterilized and turned into a liquid. It can be injected into a person's lips to enlarge them, or injected under wrinkles to fill them out. However, bovine collagen will eventually be absorbed into the body, which is why the effect lasts only for about three months.

It is not only the skin of dead cows that can be used to help plump up our lips. The filler alloderm is derived from the skin of actual dead people and can be used to create fuller lips. It is made up of thin sheets of freeze-dried human collagen that can be rolled and placed through tiny incisions to fill one lip or both. It is specially treated to reduce the risk of rejection by the body, and to ensure that it is free from disease. Alloderm has been successfully used for some time in the treatment of burns. It is also used in penis enlargement treatment, which involves several layers of alloderm strips being placed around the penis shaft to make it thicker.

Can anything live inside our skin?

Yes, there are creatures that can live inside our skin—so get ready to itch! The scabies mite is a tiny insect, too small to be seen without magnification, with a round body and eight legs. Typically, an affected human is infested by about ten to twelve adult mites. After mating, the male scabies mite dies. However, the female scabies mite burrows into the top layer of skin to set up home, and lays between one and three eggs each day. She will also leave a trail of dark-colored marks, which are basically her poo. The scabies mite's favorite locations include the hands, wrists, armpits, and genitals. The eggs and poo trigger an allergic reaction in our skin, which causes severe itching.

If you are unlucky enough to have caught "Norwegian scabies," which is more severe and highly infectious, you can expect to have thousands of the mites living in your skin. Your hands, feet, and trunk will become scaly and crusted, with innumerable live mites hiding under the crusts.

Another potential human parasite is the botfly. A month after Tanya Andrews returned from a vacation in Costa Rica, she developed a painful lump on her head. She assumed it was an abscess until it started to wriggle. It turned out to be the maggot larva of a live botfly. While Tanya was vacationing in Costa Rica, a mosquito had deposited a tiny botfly egg onto her scalp. The egg had hatched into a maggot, which burrowed under her skin and began to grow. To treat the parasite, doctors placed petroleum jelly over the lump. Maggots require air to breathe, so when the petroleum jelly cut off the supply of air it caused the maggot to suffocate and die. Once the maggot was dead the doctors used tweezers to pull it out of her scalp.

In another reported case, a Canadian woman returned home from a vacation in Peru, to find that she was developing a swelling just above her ankle. She visited a clinic in Toronto and the doctor noticed that something was moving around inside the swelling. As the doctor continued to examine the patient, he found more swellings, and in total removed eleven wriggling botfly larvae from her skin.

Is it true that books used to be bound in human skin?

Anthropodermic bibliopegy is the practice of binding books in human skin. In the eighteenth and nineteenth centuries human skin was sometimes used to bind books, usually medical books. Some doctors would specify that books they'd written should be bound in human skin, which often came from executed criminals, usually murderers. It was common to bind accounts of murder trials in the killer's own skin.

In 1821, John Horwood was publicly hanged in Bristol, England, three days after his eighteenth birthday, for the murder of Eliza Balsum. She was an older girl with whom he had become obsessed, and he had threatened to kill her. Eliza died from a fractured skull after being struck by a large stone thrown by Horwood. He was tried and sentenced to hang, and after his death his body was given to surgeons at a Bristol hospital for their dissection classes. A surgeon called Richard Smith carried out the dissection, and had Horwood's skin preserved and tanned. Horwood's skin was used to bind a book that contained the account of the murder he had carried out, the trial, and his execution.

In the 1820s, Irish immigrants William Burke and William Hare murdered many people in Edinburgh. At this time, medical science was flourishing and there was high demand for corpses. Burke and Hare apparently committed their murders so that they could make money by selling the bodies to a doctor called Dr. Knox, who used the corpses for anatomical study. Eventually, the pair were caught, but Burke was the only one prosecuted, and he was hanged in 1829. After his execution, Burke's body was donated to a medical school. Skin from his body was used to make the binding of a small notebook.

Is it possible to make furniture out of human skin?

The American serial killer Eddie Gein (1906–84), who was the inspiration for the character Buffalo Bill in the film *The Silence of the Lambs*, found a number of gruesome uses for the skin of his victims. When police entered his desolate farmhouse, suspecting Gein of committing a robbery, they made some very grim discoveries. Gein had used human skin to make an armchair, lampshade, and wastebasket, as well as a number of suits for himself. He had even made himself a human-skin belt, studded with nipples.

Another example was the horrifyingly reprehensible Ilsa Koch, the wife of a Nazi commandant at the Buchenwald concentration camp. She was a terrifying figure, who would ride around the camp half naked, and if any prisoner dared to look at her, she would have him or her severely whipped. A secret Nazi report described her as the most hated person at Buchenwald, and "a perverted, nymphomaniacal, power-mad demon." Prisoners testified to seeing items in her home that were made from prisoners' skin, such as lampshades, handbags, skin-bound books, and a pair of shoes. She also enjoyed collecting human tattoos, which had been cut from the bodies of murdered prisoners. In 1967, while in prison, Ilsa Koch was found hanging from a pipe in her cell, in an apparent suicide.

Why does body odor smell so bad?

The human body needs to keep its temperature at around 99°F (37°C), and one of the ways the body cools itself is by sweating. In hot conditions the rate of sweat production increases and heat is lost from the body as water evaporates from the skin. Sweat contains a mixture of water, salt, and toxins, and most of our sweat is produced by millions of eccrine glands, which are found all over the body.

However, we also have a different type of sweat gland, found under our arms and around our genitals, and these are called apocrine glands. Apocrine glands produce a milky sweat which contains proteins and an oily substance called sebum, which is the skin's natural moisturizer. This milky sweat is an ideal food for the many tiny bacteria that are found on the

skin under our arms and around our groins. As a result of the bacteria feeding on this sweat, they produce smelly chemicals which we smell as BO. The longer these bacteria are left to consume the contents of the sweat (in other words, the longer it takes us to have a shower), the stronger the smell will become.

Deodorants and antiperspirants work in different ways to help the body deal with sweat. Deodorants allow sweat to be released, but they contain antiseptic agents that kill the odor-causing bacteria, as well as fragrance. Antiperspirants, on the other hand, work by blocking the pores, to prevent the sweat from being released in the first place.

What is "fish odor syndrome"?

"Fish odor syndrome" is a rare, inherited condition, in which the sufferer constantly smells of rotting fish. The smell is caused by a substance called trimethylamine. In most people's bodies trimethylamine is broken down by the liver. However, this process doesn't work in the case of sufferers, so trimethylamine builds up in the blood circulation. The substance passes out of the body through the saliva, urine, vaginal fluids, and sweat, resulting in the unpleasant smell.

This condition can be triggered by a kidney or liver infection or too much intake of the chemical choline, which the body turns into trimethylamine. There is no cure, but avoiding foods that contain choline, such as saltwater fish, egg yolks, peas, liver, kidneys and legumes can help to reduce the smell. Trimethylamine is also produced by naturally occurring bacteria in the intestines.

Do we breathe through our skin?

In the film *Goldfinger*, secretary Jill Masterson is painted gold and consequently dies. James Bond comments that she died of "skin suffocation," and says that professional dancers know to leave a small area of unpainted skin at the bottom of their spines, to prevent suffocation.

In fact, we do not breathe through our skin; rather, we get oxygen from our lungs, which is then transported in the blood, so we could not die of skin suffocation. However, if our skin were to be covered in paint or anything else that would prevent the millions of sweat glands from functioning properly, our bodies would overheat. As a result of this, our body systems, including the heart and lungs, would shut down, and this would lead to death.

Is it true that we continually shed dead skin?

When cleaning and dusting around the house, you will probably be clearing up a lot of dead skin cells. Every day we shed around 10 billion skin cells, which add up to around 4½ pounds (2 kg) each year. It is thought that around 80 percent of household dust consists of dead skin cells.

The top, visible layer of your skin is called the horny layer. The cells of this layer are flat and dead, and rubbing the skin will set many of them free into the air, where they will float around and probably land on your furniture or floor.

Do moisturizers really work?

Yes and no. Moisturizers do work, but unfortunately their effects are only temporary. They work by adding water to skin cells found on the outer layer of the skin. This plumps up the skin, rather like rice cooked in water, and this makes the skin feel smoother and softer. The plumping up of these skin cells helps to reduce the appearance of fine lines, but the lines will return once the moisturizer has worn off. Some moisturizers also contain ultraviolet sunscreen to protect the skin from the sun's rays. This helps to prevent wrinkles from forming in the first place, as it is estimated that over 90 percent of wrinkles may be caused by sun exposure.

Why is smoking bad for the skin?

There are three main ways in which smoking damages the skin. First, cigarettes contain nicotine, which is a toxic substance that destroys vitamin C. In fact, nicotine is so toxic that even nonsmokers lose vitamin C when exposed to cigarette smoke. One reason why this is a problem is that our bodies use vitamin C to produce collagen fibers, which help to give strength to our skin. Collagen is the main structural protein of the skin, which keeps it elastic. When the collagen starts to break down due to the lack of vitamin C from the effects of nicotine, our skin begins to wrinkle and sag.

The second way in which smoking damages the skin is to do with our blood vessels. Blood carries vital oxygen and nutrients to the skin cells, but smoking makes our blood vessels become narrower. When the blood vessels constrict, this means less oxygen and nutrients will reach the skin cells. One cigarette may reduce the amount of oxygen reaching the skin for up to ninety minutes, and this can lead to dull, gray skin.

Third, smoking increases wrinkles by 80 percent. The continual sucking on a cigarette, and squinting through the smoke that encircles the head, leads to the formation of wrinkles.

The easiest way to think about the effects of smoking on the skin is to imagine a piece of apple left in the open, which after a while will start to turn brown. This is because of the oxygen in the air; in a process known as oxidation, the air has started to destroy the flesh of the fruit. Our skin is like the apple and the smoke is the air. Placing salt onto the apple will help prevent it turning brown so quickly. Salt works as an antioxidant on the apple and protects the flesh of the fruit. Certain vitamins such as A, C, and E will do the same for your skin, so eating plenty of fruit and vegetables is important. Nonetheless, people who have smoked for a long time will generally look ten years older than nonsmokers of the same age.

How does sunlight cause wrinkles?

Ultraviolet radiation from the sun penetrates into the deeper layers of the skin and causes damage. It causes dehydration of the skin and also interferes with the structure of collagen and elastin fibers, which leads to wrinkling of the skin. Wavy bands of tough collagen fibers restrict the extent to which the skin can be stretched, and elastin fibers return the skin back into shape after it has been stretched. Sun damage causes the collagen fibers to decrease in number, stiffen, and break apart. It also causes the elastin fibers in the skin, which are responsible for its elasticity, to lose their ability to snap back into place and, consequently, wrinkles form.

Can any skin-care products cure wrinkles?

No, but there are skin-care products that can temporarily help to improve fine lines and wrinkles, such as retinoid products and alpha hydroxyl acids (AHAs). Retinoids are chemicals that are derived from vitamin A, and AHAs are derived from fruit and milk sugars.

First, let's deal with retinoids. Our bodies are continually producing new skin cells in a process of cell division which is called mitosis. Retinoids speed up the rate at which the skin cells divide, and this helps to improve the skin's appearance. Retinoids also help to prevent the breakdown of collagen and elastin in the skin.

AHAs have been used for thousands of years as a skin-rejuvenating product. Cleopatra is reported to have bathed in sour milk, which contains lactic acid (a type of AHA), to improve her complexion. AHAs are often added to skin-care products such as moisturizers, cleansers, toners, and face masks. They work mainly as an exfoliant, by removing dead skin cells from the top layer of the skin to make room for the growth of new skin. AHAs can also promote the production of collagen and elastin.

In Elizabethan times, some women applied puppy urine to their faces, as they believed it would help to improve the health of their skin and give them a radiant complexion. The wife of the famous diarist Samuel Pepys used puppy urine on her skin, but he didn't record whether or not it had an effect!

Where do warts come from?

Have you ever noticed someone with a little cauliflower growing on his or her finger? Well, it was probably a wart. Warts are caused by the human papilloma virus (HPV), which causes the top layer of skin to grow too much. Contrary to popular belief, a wart does not contain a "root." A wart may contain black dots, but these are in fact blood vessels, which contain blood that has clotted.

Warts usually affect the hands, but they can also affect the feet, in which case they are called verrucas. They can be caught by touching another infected person, or, especially for verrucas, by walking around a swimming pool or changing room where an infected person has recently been. They are easier to catch if the skin is broken.

At one time it was thought that rubbing a dusty toad onto the wart, or rubbing a cut potato onto it and throwing it over a fence, would help to get rid of the wart. However, warts usually disappear by themselves, although they may take months or even years to go away.

Why does the skin on our fingertips wrinkle during a bath?

Our skin is made up of layer upon layer of skin cells. New cells are constantly growing from below, pushing older cells toward the surface. As skin cells reach the surface, they get drier and thinner. These dead cells lie in overlapping stacks on the surface of the skin, until they flake off or are rubbed off.

Our fingertips, palms, and the soles of our feet have thicker layers of dead skin cells than the rest of our bodies. When we take a bath these dead cells begin to plump up, because they absorb some of the water. If you stay in the bath for a long time, these skin cells will continue to expand, until they are no longer able to lie flat next to each other. As a result, they begin to fold and overlap, causing the skin to become wrinkled.

Is acne contagious?

Acne is a skin condition that typically causes one or more of the following: red or yellow spots, blackheads, whiteheads and greasy skin. It often affects the skin of the face, back, neck, and chest.

Acne is not contagious, so will not transfer from one person to another, but if the acne sufferer scratches the spots, this may cause the bacteria to spread below the spot, which will cause further inflammation, leading to more redness.

In the 1500s many acne sufferers applied a lotion to their

17

skin called Soliman's Water, which was used to help get rid of spots and warts. It was said to have been quite effective, which was probably due to the fact that it contained mercury, which burned off the skin, and also caused gums to recede and teeth to fall out! Exposure to high concentrations of mercury vapor caused harmful effects on the nervous, digestive, and respiratory systems, and the kidneys.

Why does a blackhead have a black head?

Most of our body is covered in hair, which grows out from our pores. Inside a pore, there is a pit called the follicle. Like the top layer of skin, which is continually shedding dead skin cells, our follicles are also partly lined with dead skin cells. Follicles also contain something called a sebaceous gland, which produces a substance called sebum—an oily substance which helps to moisturise the skin to keep it smooth and supple. When the body produces too much sebum, it mixes with the dead skin cells inside the pore and forms a plug inside the follicle. The sebum and dead skin cells continue to build up and solidify as a soft white substance. If the surface of the pore is covered by skin, it is called a whitehead. If the pore is open, without any skin covering, the top of the plug is exposed to the air. The oxygen in the air reacts with the substance inside the pore and causes it to darken, resulting in a blackhead.

How does "fake tan" work?

The first self-tan products came out in the 1960s, and when applied to the skin, they gave people a bright orange glow. Thankfully, things have moved on, and there is now an array of products that give the skin a more natural-looking tan.

Our skin is made up of two main layers: the epidermis, which is on the outside and is the part we can see, and the dermis, which is found below it. The epidermis is itself composed of layers. The deepest layer is called the basal layer, and it is this layer that is affected by real sun tanning. The outermost layer of the epidermis is called the horny layer, and it is this layer that we can see and touch. Most self-tan products affect only this horny layer rather than the underlying basal layer.

The most effective fake tan products are those that contain dihydroxyacetone (DHA) as the active ingredient. DHA is a colorless sugar that reacts with the dead skin cells of the horny layer. As the sugar reacts with these skin cells a color change occurs, which usually lasts about five to seven days. Products that create a darker tan usually contain more DHA than those that produce a light tan.

Every day, millions of our dead skin cells are rubbed off or worn away from the surface of the skin. This is why fake tans gradually fade; as the dead cells of the horny layer are worn away, so is the tan. Consequently, these products need to be applied regularly for the tan to be maintained.

Can the smell of sweat be a turn-on?

Apocrine sweat glands are mostly found in our armpits and around our groins. These glands become active at puberty and release a milky sweat, which is regulated by our hormones. This milky substance contains smelly chemicals called phero-mones, which are believed to send out subconscious scent signals thought to trigger sexual responses. You can even buy colognes containing human pheromones to help increase your animal magnetism!

Scientists have long known that pheromones help animals to attract each other, but it is less clear how they affect humans. Studies have suggested that women can detect cer-tain odors in sweat, which help them to select males who will increase chances for healthy offspring. It is believed that the pheromones found within sweat play a part in the attraction. Interestingly, research shows that pheromones are not actu-ally produced by the sweat glands; instead, they result from the action of bacteria on the person's sweat. Therefore, if we didn't have bacteria in our sweat, we would not be able to produce pheromones.

In 1995 Switzerland's Bern University carried out an exper-iment to test the power of pheromones. A group of women at the midpoint of their menstrual cycle, which is when their sense of smell would be keenest, were asked to smell some un-washed T-shirts worn by different men, some of whom were their relatives. The men had worn the T-shirts to bed for two nights, and the women were asked to sniff the shirts and reject any shirts they didn't like the smell of. The majority of the women rejected the shirts worn by males closely related to them and consistently preferred shirts worn by men whose genes were most unlike their own. On this basis, the researchers at the university concluded this difference in genes would mean the offspring would probably be healthier than children born of people who were related. When two blood relatives have a child together, there is a greater risk of its suffering with an inherited disease. Therefore, having a child with a partner who is genetically very different will mean a smaller chance of the child's inheriting a disease, so the child is likely to be healthier. This research was repeated in America and Brazil with similar results.

Interestingly, it has also been suggested that birth control pills may reverse a woman's natural preference in men, so that she becomes attracted to men who would not normally appeal to her—meaning that she could possibly reject a suitable partner.

What causes dark circles under the eyes?

The presence of dark circles under the eyes can often be an inherited characteristic. The skin under our eyes is very thin, and when blood passes through the veins near the surface of the skin, it can produce a bluish color. Some of us inherit skin that is quite thin and transparent, and as a result the area around our eyes may appear quite dark. If this is the case, there may be an Uncle Fester somewhere in the family.

Dark circles can also be caused by lack of sleep, as excessive tiredness can cause paleness of the skin, which again allows the blood underneath the skin to become more visible, making the skin under the eyes look darker. Skin can also be pale due to a lack of iron, perhaps caused by menstruation or pregnancy, and this too will make the dark circles more visible.

How can skin cells be used to help solve crimes?

A standard method for identifying a person is to take a copy of his or her fingerprints using ink and paper. Our fingerprints are unique and remain the same throughout our lives, so they can therefore be used to distinguish one person from another. Similarly, the DNA (deoxyribonucleic acid) of almost every individual is unique. DNA is found in most body cells, such as skin and blood cells, and includes all the information needed to make a person. It is made up of lots of chemicals, which have many different possible combinations. With the exception of identical twins, no two people have the same DNA. In paternity suits involving identical twins, if neither brother has an alibi to prove that he could not have impregnated the mother, the courts have been known to force them to split child-care costs.

A British geneticist coined the term "DNA fingerprinting" in 1985. Actually, it is probably more appropriate to call the method "DNA typing," since it has nothing to do with fingerprints. DNA typing involves taking a sample of skin, hair, blood, semen, or saliva, and extracting DNA from this sample. The DNA is placed into a device that separates it into its constituent parts and displays this pattern as a column made up of dark and light bands, rather like a bar code. The pattern of these bands varies greatly from one person to another. Therefore, this DNA evidence can be used either to convict the guilty or exonerate those wrongly accused.

When skin on our fingertips grows back after an injury, does the fingerprint have a new pattern?

The swirling, looping pattern found on the skin of our fingertips stays with us for life. When an injured fingertip heals, the regrowth of skin will show the same pattern as before.

In the 1930s, the famous American bank robber John Dillinger decided to change his fingerprints to avoid prosecution. He used acid to burn the ridge patterns of his fingertips and assumed the fingerprints were gone. When his new skin grew back it kept the old pattern, so he had the same finger-prints as before.

In 1934 the FBI was alerted by an informant that Dillinger was inside a theater. As he walked out, the FBI agents closed in and ordered him to put up his hands and surrender. According to the agents, Dillinger reached into his pockets as if going for his gun, so they shot him dead. Almost immediately, people came forward to collect souvenirs; women even dipped hand-kerchiefs in his blood. Dillinger was taken to the morgue and was thoroughly examined. The FBI agents took his finger-prints and were able to match them to the fingerprints they had previously taken.

In Miami in 1990, police arrested a man whom they be-lieved to be a drug dealer. When they fingerprinted the suspect they found his fingerprints consisted of strange zigzag pat-terns. They concluded that the man had sliced the skin on his fingertips into tiny pieces, and then transplanted those pieces onto his other fingers. After the fingertips had healed, his new fingerprints had unusual patterns. The police came up with a brilliant, creative solution. They took photographs of his fin-gerprints, cut them up into small pieces, and then rearranged them like a jigsaw puzzle. Using this technique, they managed to re-create his original fingerprints, and he was consequently convicted.

How do lie detector tests work?

Lie detector tests use something called a polygraph machine to measure the body's physiological responses. The underlying theory is that when people tell lies, they also become nervous about lying. If someone is lying, it is supposed to trigger certain changes in the body. For example, when we lie we tend to produce more sweat, our breathing rate changes, and there are changes to our heart rate and blood pressure.

The average person has around 2.6 million sweat glands in his or her skin, and the hands contain thousands of them. A polygraph machine measures the galvanic skin resistance, which is basically a measurement of the sweat on our fingertips. Fingerplates are attached to two of the subject's fingers, and these plates measure the skin's ability to conduct electricity. When the skin is sweaty, it conducts electricity much more easily than when it is dry. Therefore, a person will generally conduct electricity more easily when lying than when telling the truth.

In the Middle Ages, judges devised a rather less sophisticated lie detector test. Officials would put flour into a suspect's mouth. If he or she was innocent, they believed, the suspect would produce enough saliva to enable him or her to swallow the flour. However, if his or her mouth was too dry and he or she was unable to swallow the flour, he or she was considered to be guilty.

Why are some babies yellow?

The medical condition jaundice can turn the sufferer's skin yellow, and sometimes also the whites of his or her eyes. This is because of a substance called bile, which is stored in our gallbladder. Bile serves a useful purpose: after we eat food, it helps to break down fats into small droplets. Bile contains a yellow substance called bilirubin. If the liver is diseased or the bile ducts are blocked, bilirubin gradually builds up in the blood, and affects the skin and the whites of the eyes turning them the characteristic yellow color.

Jaundice often affects newborn babies during their first few weeks of life. This is because their livers take a while to get working properly.

Can fingernails simply fall off?

Like bacteria, fungi are found all around us. They usually become a problem only at certain times; for example, if our immune system is compromised or we have a cut on our skin.

Fungal infections can affect one or many nails and are often painless to begin with. As the infection grows, the fungus spreads from the tip of the nail toward the bottom of the nail. Sometimes, fungal infections are mild, causing thickening and slight discoloration of the nail. More severe fungal infections can cause nails to become yellow, green, or black, and white patches can form on nails that have come apart from the skin. Some fungal infections can become so severe that the nails may even start to crumble and fall off.

What can we tell about a person from the state of his or her nails?

The color and condition of a person's nails can give clues as to his or her state of health:

- If some or all of the nail is white in color, it could be a sign of fungal growth or liver problems.

- White dots may indicate a zinc deficiency or a slight injury to the nail.

- Brown nails may indicate fungal growth or kidney disease.

- Yellow nails could be due to jaundice.

- Blue nails can indicate poor circulation or a heart or lung condition.

- Pitted nails may suggest a skin condition, such as psoriasis or eczema.

- Nails that break or split easily could indicate a thyroid problem.

- Nails that have a yellowish tint and pink coloring at the base may be linked to diabetes.

- Curved nails may indicate a respiratory disorder.

Can maggots be used to clean wounds?

Placing maggots on a wound may sound rather unhygienic, but it has been found to be an extremely successful way of cleaning infected wounds for patients who have not responded to conventional treatments. The arrival of antibiotics in the twentieth century meant the use of maggots fell out of favor, but it is now making a comeback and is used today in some hospitals in the U.K. Maggots are currently used for treating conditions such as leg ulcers, pressure sores, and other infected surgical wounds.

How, you may be wondering, could maggots possibly help? Well, when they are introduced to the wound, the sterile maggots feed on the dead flesh but leave behind the healthy flesh, and in this way they clean wounds in a fraction of the time taken by other types of treatment. The maggots also produce helpful chemicals that kill some of the bacteria that the maggots themselves don't consume.

To produce the necessary sterile maggots, flies are kept in a sealed room and fed meat, onto which they lay their eggs. These eggs are separated and sterilized, and they then develop into maggots, which are ready and willing to munch on dead flesh!

Why do certain people attract mosquitoes?

Like little vampires, mosquitoes suck blood from their victims, but it is only the female mosquito that feeds on humans. Since male mosquitoes do not suck blood they do not transmit diseases, and prefer to sip on flower nectar and plant juice. The female mosquito is usually attracted to a person by his or her smell and temperature, but the mosquito may also be attracted by somebody's looks. It seems that mosquitoes may prefer blonds to brunets, but this might simply be because blonds are more noticeable to them than brunets.

When a mosquito is attracted to someone, it will jump onto his or her skin, inject a small amount of saliva, and then begin drawing blood. The saliva helps the mosquito to penetrate the skin more easily, and also stops the blood from clotting, making feasting on the blood easier. The bump or welt that occurs after the mosquito attack is an allergic reaction to this saliva. It's best to avoid scratching a mosquito bite, as this only serves to spread the saliva deeper into the skin.

Every year, a million people die because of diseases transmitted by mosquitoes, such as malaria. However, mosquitoes are unable to transmit HIV, the virus that causes AIDS, because this virus cannot live inside a mosquito. Even if it could, the amount of the virus carried would be too small.

Like vampires, mosquitoes hate garlic, so eating garlic will help to keep them away.

CHAPTER TWO

Follicular Fancies

Is baldness in men a sign of virility?

Some experts believe that baldness may have evolved as a way for men to show women that they are hormonally sound and ready to mate. Male pattern baldness is passed down in the genes and is caused by the hair follicle becoming more sensitive to male sex hormones, such as testosterone. This causes the hair follicle to shrink, which stops the hair from growing. Male hormones circulate in the blood and are present in high levels after puberty. Therefore, balding could possibly be a nonverbal signal to tell women, "I've got testosterone and I'm ready to reproduce, baby!" Unfortunately, most women today do not seem to interpret baldness in this way.

A few historical figures were particularly touchy about being bald, such as Paul I, Tsar of Russia (1754–1801). He imposed a ban on words such as "baldy" and "snub nose," and if anyone mentioned his baldness he or she would be flogged to death. A soldier who unfortunately made a comment about the tsar's lack of hair was executed.

Do bugs live in eyelashes?

Most people don't like the idea that bugs can live on their skin and hair. However, the truth is that many bugs do, and they live with us in harmony, most of the time. By the time we reach late adulthood, most of us have wiggly, microscopic, wormlike mites called demodex mites living in the roots of our eyelashes. If you pull out one of your eyelashes and examine it under a strong magnifying glass, or better, a microscope, there is a good chance you will see one of these tiny mites clinging to the base of the lash. They can also live in our skin pores and the hair follicles on our face, such as the eyebrows.

These mites are cigar shaped, a third of a millimeter (a tiny fraction of an inch) long, and have eight stubby little legs situated at the front of their bodies, so they waddle along fairly slowly. When one of these mites reaches a hair, it burrows headfirst down into the follicle. Their bodies are layered with scales, which help to anchor them into the follicle, and their needlelike mouths eat dead skin and oil that is produced by the skin. Fortunately, although the mites eat, they don't actually poo in the follicles.

An individual female can lay a number of eggs in a single follicle. When mature, the mites leave the follicle, mate, and find a new follicle into which they lay their eggs. Each mite can live for several weeks, and mites can be transferred between humans if two people's hair, eyebrows, or the sebaceous glands on their noses come into close contact.

Mites living on our eyelashes are usually quite harmless, and most people are totally unaware of the little squatters living in their hair follicles. However, if too many accumulate in a single hair follicle, they can cause itching, certain skin disorders, or an eyelash to fall out. As many as twenty-five eyelash mites have been found huddled together in a single follicle! There are some great pictures of these mites in a variety of poses on the Internet.

Is there such a thing as a "bearded lady"?

A small number of women are able to grow enough facial hair to have a beard. The world record for the woman with the longest beard is held by an American lady called Vivian Wheeler, whose beard measured 11 inches (28 cm) in 2003. She has been growing it since 1993.

So how is it that some women can display such an apparently masculine trait? Well, it all comes down to hormones. Both men and women produce male and female hormones, but male hormones, which are also called androgens, are found in larger amounts in men. Female hormones help to stunt the growth of facial and body hair, but if a woman produces too many androgens, such as testosterone, it can lead to excessive hair growth (also known as hirsutism) which may result in her growing a moustache and beard.

Levels of androgens in the body can increase during puberty, pregnancy, menopause, and at times of stress, so all of these factors could lead to excessive hair growth. Medical conditions such as ovarian cysts and anorexia nervosa are also potential causes.

There have been a number of famous bearded women down the years, including a Mexican-born lady called Julia Pastrana (1834–60). She made a living by performing all over the world as the "Bearded and Hairy Lady." She had a condition called hypertrichosis, which causes an excessive growth of hair on the body. She also suffered from swollen gums and a jutting jaw. In 1860 she gave birth to a boy who had inherited the same condition. He died three days later, and Julia died soon after. Her husband and promoter, Theodore Lent, had the bodies of his late wife and child mummified, and for many years he exhibited their corpses in a glass case in his traveling show.

Why do some people have ginger hair?

People who have red hair, fair skin, and freckles have inherited their coloring from their parents, even though the parents themselves might not be redheads. Some scientists believe the gene responsible for ginger hair may have originated in Neanderthal Man, who lived in Europe for 260,000 years before the ancestors of modern man arrived from Africa about 40,000 years ago.

A substance called melanin is responsible for determining our hair color. There are two types of melanin, which are called eumelanin and pheomelanin. Eumelanin is the pigment we associate with a suntan, which helps to protect our skin against UV rays. In the hair, eumelanin produces a black color. Ginger hair and brown hair, on the other hand, are caused by pheomelanin. Red-haired people produce more pheomelanin than eumelanin, which is why their hair is red, and this is also why they tend to have paler skin and often suffer badly from sunburn. Scotland has the highest proportion of redheads of any country in the world, with natural redheads making up around 13 percent of the population.

Are there health benefits to having ginger hair?

Scientists believe that redheads are able to fight off certain debilitating and potentially fatal illnesses more effectively than blonds or brunets. The reason for this is that pale complexions allow more sunlight into the skin. As a result, more of the sun's UV rays penetrate through the skin's layers, encouraging the production of vitamin D in the body, which helps to prevent a range of conditions including rickets, and the lung disease tuberculosis, which can be fatal. Even as recently as fifty years ago, rickets was a fairly common disorder, which led to the weakening and bowing of people's bones.

However, there are also downsides to having ginger hair. For one thing, people with pale skin have an increased risk of skin cancer if exposed to the sun's rays for long periods.

What are eyebrows for?

Eyebrows help to keep water out of our eyes when we sweat or walk in the rain. The arched shape of the eyebrow helps to redirect the rain or sweat to the sides of our face, keeping our eyes relatively dry. Without eyebrows, walking in the rain would be rather more uncomfortable. Also, diverting sweat away is helpful because the salt in our sweat can irritate our eyes, making them sting.

Our eyebrows are also useful for expressing emotion. Eyebrows can become very animated when we feel emotions such as anger or surprise.

Wearing animal skins as clothing has often been fashionable, but in the 1700s it was trendy for high-class men and women to wear fake eyebrows made from mouse skin. First, they would trap a mouse and then skin it. The skin was cleaned and eyebrow shapes were cut out from it. These trendy folk then shaved off their own natural eyebrows and used glue, made from fish's skin and bones, to fix the mouse's skin onto their brows.

What happened to *Mona Lisa*'s eyebrows?

In Leonardo da Vinci's painting, *Mona Lisa* has a famously enigmatic smile, which has provoked debate for centuries. However, *Mona Lisa* is unusual in another way, which has been relatively overlooked—she has no eyebrows. Da Vinci began the picture in 1503, and it took him about four years to complete. Da Vinci painted his model, who is believed to have been Lisa Gherardini, when she was twenty-four years old. Mona was not her first name but is short for Madonna, which in this context means "Italian Lady." Some researchers claim that it was common for ladies of this era to pluck out their eyebrow hairs, as this was considered fashionable and attractive—hence the absence of eyebrows. Other experts think the eyebrows were omitted because da Vinci simply hadn't finished the picture.

In 1911, *Mona Lisa* was stolen from the Louvre, in France. It was recovered two years later, but it is said that during this hiatus more people visited the empty space where *Mona Lisa* had originally been positioned than had previously visited the actual painting!

What is pubic hair for?

There are a number of different theories as to the purpose of pubic hair. The most likely explanation relates to pheromones, which are scents produced by the body that are believed to stimulate sexual arousal in potential partners.

Our bodies contain apocrine sweat glands in our armpits, pubic region, and other areas. These glands release a milky type of sweat, which contains pheromones. Some scientists believe that our pubic and armpit hair is designed to trap these erotic scents, for the purpose of increasing the pheromones' effectiveness.

However, there are a number of alternative theories. Some experts argue that pubic hair provides cushioning for the genitals, helping to prevent friction during sex. Anthropologists, on the other hand, suggest that in ancient times a man's pubic hair may have had the function of impressing or discouraging any male competition—rather like a lion's mane!

Pubic hair is curly because of the shape of the hair follicles, which are the pits in which the hairs sit. Our pubic hair follicles have a flattish, oval shape, which cause the hair to bend as it grows, resulting in curly, twisted hairs. Other parts of our body may have round follicles, which produce straight hair.

Did Victorian women have bikini waxes?

In Europe in Victorian times, the removal of pubic hair was unheard of. Most Victorian women tended to let it grow naturally and didn't attempt to style or cut it. During this prudish era, pubic hair was a taboo subject which, if mentioned, could cause great embarrassment and offense. Most Victorian nude paintings hide the subject's genitals, and those that showed women in explicit poses tended to paint them without pubic hair. It is said that the English art critic John Ruskin (1819–1900) had been so used to seeing hairless females on canvases that when he first saw his wife's pubic hair, he was so horrified that he never consummated his marriage.

The first commercial for a female hair removal product came in 1915, when *Harper's Bazaar* printed an ad showing a woman in a sleeveless evening gown with hairless armpits. At the same time, Wilkinson Sword launched a marketing campaign aimed at women. Within two years, sales of razor blades had doubled.

In the 1950s, the bikini was invented, and the bikini wax followed soon after.

What is the difference between Brazilian and Hollywood bikini waxing?

One thing is for sure, both types involve the kind of medieval pain once reserved for the village witch. The term "Brazilian waxing" derives from the special kind of bikini wax favored by many Brazilian women, which allows them to feel confident parading down the beach in their dental-floss bikini bottoms. This waxing treatment traditionally leaves a small rectangle of hair, or "landing strip," on the mons pubis, the area found just above the genitals. The Brazilian wax usually includes the waxing of the labia as well as between the buttocks (ouch!), which is why it can be more painful than a traditional bikini wax.

However, those Hollywood gals had to take it one step further, and created the Hollywood bikini wax. This involves whisking the whole lot off, leaving it totally bare down there. Some women have claimed that a "Hollywood" is even more painful than giving birth, so it may be best to have a few stiff drinks before attempting this one.

How can a hair sample be tested for drugs?

Analysis of human hair can help to provide evidence of the use of illegal drugs. This analysis can also reveal the extent and timing of drug use, by testing different parts of the hair. Even if someone's hair has been colored or permed, traces of any drugs taken will still remain.

Here's how it works: our hair needs oxygen and nutrients (food) to grow, and these are carried to the hair follicles by our blood. Blood is also the means by which drugs travel around the body, and, consequently, small traces of any drugs present in the blood can become trapped in the follicle. As the hair continues to grow, from the bottom of the follicle, it will contain tiny traces of the drug, which cannot be washed out.

Scalp hair grows at a rate of about ½ inch (1 cm) each month, so it's even possible to measure how recently drugs were taken, simply by testing different sections of the hair.

Oops.

Have beards ever been taxed?

Modern Russia started with the rule of Peter the Great (1672–1725). His original title was simply Peter I, but he declared himself Peter the Great in 1721. No one was likely to contradict him, for he was 6 feet 8 inches (2 m) tall and very strong. He believed that the way forward for Russia was to become Westernized, and because European men were usually clean-shaven, he introduced a yearly beard tax of 100 rubles (although priests, peasants, and women were exempt). He later added a penalty that involved shaving with a blunt razor. Peter came up with many other unusual tax schemes including taxes on births, marriages, burials, salt, hats, beehives, beds, firewood, and drinking water.

Why do some parts of the body have thicker, longer hair than others?

Scientists believe that our hairs are programmed to grow to a certain length. There are good reasons for this; it would be impractical, for example, if we were to grow extremely long eyelashes. However, as we get older, some of our hairs can forget their programming, which is why older men (and some women) develop bushy eyebrows and long ear hair.

Hair growth can also be affected by the way in which we remove it. If, for example, we remove hair from a particular area many times, we may damage the root, thus preventing regrowth. This is why treatments such as tweezing and waxing can lead to the permanent loss of hair.

What would Shakespeare's Juliet have looked like?

William Shakespeare (1564–1616) lived during the reign of Elizabeth I. In those days, the archetypal beautiful woman would have had light-colored hair and a pale, white face with red lips and rosy cheeks. A pale complexion was a sign of wealth and nobility, as poor women would generally work outdoors and therefore have darker skins. It was fashionable for Elizabethan women to dye their hair yellow, using plant extracts such as saffron, celandine, and cumin seeds. They used lipstick and blusher, which were made of insects called cochineals, and painted their faces with white makeup, which contained lead and vinegar. Because the lead was poisonous, it would often make their hair fall out over time—fortunately, it also became fashionable for women to have high foreheads, where their hair had fallen out due to effects of the lead. However, the lead poisoning could have more dramatic consequences, including death.

The reality in Shakespeare's day was that most people were poor, washed rarely, and suffered from bad nutrition. Their hair would probably have been dry, filthy, and filled with lice. They often suffered with skin conditions such as eczema, ulcers, scabies, and other skin infections. They had little, if any, dental care, so people often had rotten teeth and bad breath. In Shakespeare's day, women were not allowed to act on the stage, so the role of Juliet would have been played by a young boy, wearing white makeup. This makeup was of course lead-based, so as a result many of the young actors were very unhealthy, with unpleasant facial skin diseases.

Does anything eat facial hair?

There are bugs that will eat just about anything, including facial hair. The American cockroach, *Periplaneta americana*, is a large, brown, winged cockroach, about 1½ inches (4 cm) long. This cockroach is commonly found in the southern United

States in tropical climates, and will often be found living in sewers. It will eat practically anything including leather, book-bindings, glue, flakes of dead skin, and soiled clothing. It has also been known to munch on the eyelashes, eyebrows, finger-nails, and even toenails of people while they are asleep.

Are fetuses covered in hair?

Around the fourth month of pregnancy, the human fetus grows a mustache of very fine hair. Over the next few weeks, the fetus will grow hair all over its body, until by the fifth month the fetus is completely hairy. This hair is known as lanugo; it is an unpigmented, fine, silky hair that is usually shed at around eight months and then swallowed by the baby. Meconium is the name given to the baby's first poo, and it is a greenish-black substance composed of materials ingested during its time in the uterus, including hair, mucus, amniotic fluid, bile, and water. Meconium is sterile and has no odor.

A rare inherited condition called congential hypertrichosis lanuginose is characterised by excessive hair growth on a child at birth, in which most of the body is covered with lanugo hair. The hair continues to grow, and this excessively long, fine hair persists throughout life.

There is a similar condition called congenital hypertrichosis terminalis, which also causes all-over body hair growth. But this hair is colored, and it is thicker than lanugo hair. Two famous sufferers of this condition are Danny and Larry Ramo Gomez from Mexico, who are also known as "The Wolf Boys." They have excessive dark hair growth that covers most of their bodies. This condition is almost always associated with a condition called gingival hyperplasia, which is characterized by enlargement of the gums. As a result, people with this condition often possess very few teeth.

Can bugs live in wigs?

In the 1600s, the English diarist Samuel Pepys was rather perturbed when he found nits in his new wig. Pepys noted on 27 March 1667:

> I did go to the Swan; and there sent for Jervas my old periwig-maker and he did bring me a periwig; but it was full of nits, so as I was troubled to see it (it being his own fault) and did send him to make it clean.

Marie Antoinette (1755–93) was the queen of France, until she was beheaded in 1793. She liked to wear extravagant clothing and made it fashionable to wear very big wigs, some of which were so gigantic that the wearer would be unable to sit upright in her carriage and would have to lie down. Women would wear these wigs for months without cleaning them and used pomade to help hide any unpleasant smells. It wasn't unusual to find bugs, or even small rodents, crawling around inside these wigs. To help get rid of lice and any other unwanted pests, the wigs were occasionally boiled or put into an oven.

What is a merkin?

A merkin is a pubic wig, an item whose origins are believed to date back to around 1450.

In the 1600s, merkins were frequently worn by prostitutes to help cover up evidence of sexually transmitted diseases. Before attaching the merkin, the women would shave off their own pubic hair, and this had a further benefit, as it helped to get rid of pubic lice, which could cause intense itching. In the days before antibiotics, it was common for prostitutes to become infected with sexually transmitted diseases such as syphilis and gonorrhea. The first symptom of syphilis is a painless open sore, which usually appears around the vagina or on the penis. Gonorrhea is another sexually transmitted disease, which also affects the genitals of both sexes. It causes itching, a yellow or green discharge, and burning or difficulty with urination. Merkins would cover up the symptoms of both diseases and were thus a boon to prostitutes of the time.

How can hair kill a human?

Some people have the unfortunate habit of chewing their hair and swallowing it. This can lead to clumps of swallowed hair becoming stuck in the stomach or intestines and hardening to form what is known as a trichobezoar (or bezoar for short), which can cause obstructions, bleeding, and perforations.

In 1999, a British girl tragically died after an operation to remove a giant bezoar, the result of eating large quantities of her own hair, from her stomach. She was rushed to the hospital after complaining of stomach pains, but while recovering from the surgery she suffered internal bleeding and died. The hairball was 12 inches (30 cm) long, 10 inches (25 cm) wide, and 4 inches (10 cm) thick. It was the size and shape of a rugby ball, and it filled her entire stomach.

A bezoar can be made up of a ball of food, mucus, hair, vegetable fiber, and/or other matter that cannot be digested by the body. In the book *Harry Potter and the Half-Blood Prince*, Harry uses a bezoar derived from a goat's stomach to save Ron's life after he has been poisoned. In the 1600s, it wasn't uncommon for people of high status to be poisoned, mostly through the use of arsenic-laced drinks, and so they found ways to help protect themselves against the threat. They commonly used animal bezoars taken from the stomachs and intestines of goats, deer, and sheep. This type of bezoar, which would have resembled a small stone, would be dropped into the person's drink to remove any toxins. Modern research has shown that bezoars can be effective in removing certain poisons such as arsenic when used in this way, as the toxic constituents found in arsenic stick to sulphur compounds found in the hair proteins in a bezoar.

Were the ancient Greeks hairy?

The ancient Greeks hated body hair and liked their bodies to be completely hairless, as they thought hairlessness represented youth and beauty. This is reflected in Greek male statues, which do not show any chest hair, and female statues, which do not display any pubic hair. Greek statues also often show men with very small penises. It is believed that ancient Greeks thought that large genitals reduced sexual potency.

The ancient Greeks are believed to have devised a way to remove hair, similar to waxing, using a substance consisting of ivy gum extract, bat's blood, powdered viper, and animal fat. It was applied to the hairy body part with a strip of linen, then ripped off.

Why did the ancient Egyptians shave their heads?

In ancient Egypt, it was common for both men and women to shave their heads. They did this to keep themselves cool and also to avoid lice infestation. At first they shaved their heads by using a stone blade, but later Egyptians used a bronze razor. Queen Nefertiti of Egypt, renowned at the time as the most beautiful woman in the world, was as bald as a coot.

The Egyptians took great pride in their appearance and liked to wear wigs. The expensive wigs worn by rich people were made of real hair, but the poor had to make do with wigs made of palm leaves, straw, or sheep's wool. In ancient Egypt, it was common practice to remove body hair too, as its presence indicated that you were a slave.

Does our hair continue to grow after we die?

In order to grow, our hair and fingernails require oxygen and nutrients, which are both carried in the bloodstream. Of course, after death the heart stops working, so blood is no longer pumped around the body—consequently, the hair and nails stop growing. It is true, however, that hair and fingernails do often appear longer after a person has died. This is not because they have grown but rather because our skin shrinks after death, making the hair and nails appear longer.

Has it ever been fashionable to wear false beards?

In the 1300s, it became very fashionable for Spanish men to wear long black false beards, and as a result, hair became extremely expensive. However, this fashion created a social problem, as everyone now looked the same, so criminals got away with crimes and innocent people were being imprisoned. Finally, King Peter of Aragon stepped in, and a law was passed forbidding the wearing of false beards in Spain.

Howard Carter (1874–1939) was a British archaeologist who, in 1922, famously discovered the tomb of King Tutankhamen in Egypt. In 1902, while supervising excavations in the Valley of the Kings, he discovered the tomb of Hatshepsut. Born in the fifteenth century B.C., Hatshepsut was one of the few female pharaohs to have reigned in Egypt. The Egyptians didn't think a woman could be pharaoh, so Hatshepsut pretended to be a man. She dressed in the traditional male pharaoh attire, which included a false beard, and fought with her army. She ruled for about fifteen years, until her death in 1458 B.C.

Why did Bjorn Borg favor "designer stubble"?

Swedish tennis star Bjorn Borg is renowned for his designer stubble, but this wasn't purely a fashion statement. Borg was extremely superstitious, and he believed that his stubble brought him good luck and that he wouldn't win Wimbledon if he shaved before it started. Perhaps it worked, as he won Wimbledon for five years running, until the curly haired but clean-shaven John McEnroe won it in 1981.

Borg comes from a superstitious family. In 1979, Borg's grandfather, while fishing and listening to the French Open final on the radio, spat into the water. At that instant, Borg won a point against Victor Pecci of Paraguay. Believing that the spit had helped Borg to win the point, Borg's grandfather continued spitting throughout the match, and went home with a sore throat. Borg won in four sets.

Did Iron Age men use hair gel?

In 2003, well-preserved human remains were discovered in peat bogs in central Ireland. These Iron Age men had lived more than 2,000 years ago and were thought to have been wealthy individuals, as it was evident they did not carry out manual work. Their nails and hair were so well preserved that scientists were able to observe that one of the men had manicured nails and the other wore hair gel! The gel was made of vegetable oil mixed with resin from pine trees and was thought to have been imported from France. He had used the hair gel to create a Mohican-style hairdo.

Unfortunately, although both men appear to have been aristocrats, there was evidence that they had been tortured and killed in their early twenties. The man with the hair gel had suffered ax blows to the head and chest and had also been disemboweled. The other man had been stabbed, his nipples had been sliced off and holes had been cut into his upper arms, through which a rope was threaded. This was probably to ensure he could not get away. Finally, he was beheaded. It is thought they were killed either as a punishment or as human sacrifices to pagan gods.

CHAPTER THREE

Skeletal Singularities

Could drilling a hole in your skull help to relieve depression?

Trepanation is a medical practice that essentially means cutting into the skull. It goes back as far as Stone Age times—archaeologists have found trepanned skulls dating back to 3000 B.C. In the Stone Age era, trepanation was carried out using sharpened stones. After the procedure, pieces of the removed skull were often collected, so that they could be shaped and polished and worn as a protection against disease.

Ancient Egyptians used a mallet and chisel to practice trepanation. They believed that the procedure helped to alleviate pressure on the brain, such as the pressure caused by bleeding from a severe blow to the head. The practice continued into the Middle Ages, when it was thought that trepanation would help to release evil spirits from the heads of the possessed.

Even today this procedure is carried out in some countries

to help treat conditions ranging from depression to epilepsy. The founder of modern trepanation is a Dutch man by the name of Dr. Bart Hughes. In 1962, he became convinced that the amount of blood in the brain controls our state of consciousness. Dr. Hughes believed that humans had been deprived of an elevated state of consciousness as a result of our evolution to walking upright, which puts the heart below the brain. Therefore, he argued, trepanation would help people to achieve this higher state of consciousness, by allowing blood to flow more easily around the brain, thus increasing our alertness and concentration.

In 1965, after years of experimentation, Dr. Hughes bored a hole into his own skull, using an electric drill, a scalpel, and a hypodermic needle (to administer a local anesthetic). He immediately began advocating the benefits of his new state of consciousness. People who have been trepanned report an increased ability to concentrate, and generally say they feel good for a long while afterward.

An English woman called Heather Perry learned about trepanning after exchanging e-mails with a man who had himself undergone this operation. In an attempt to cure chronic fatigue syndrome, she traveled to America to meet

him, and with his help carried out this procedure on herself. The operation was filmed and broadcast on national U.S. news. First, she injected a local anesthetic into her head, then she used a surgeon's knife to cut away a section of her scalp before drilling a ¾ inch (2 cm) hole into her skull. Unfortunately, she drilled too far and severed a membrane protecting her brain, and so required urgent medical attention. However, afterward she said she had no regrets and claimed that her health was much improved. She said she generally felt better and had more mental clarity.

Nonetheless, surgeons point out that there are serious risks involved with trepanation, such as blood clots, infection, and brain injury. It is also a painful procedure—although the brain itself does not feel pain, cutting through the skull, muscle, and skin would be very painful. There is also, of course, no serious evidence to suggest it actually works.

Why do our knuckles click?

Many people find the noise associated with cracking joints rather unpleasant and believe the sound is caused by bones rubbing together. In fact, this cracking noise isn't due to the rubbing of bones but is caused by gases between the joints.

Our joints contain something called synovial fluid, which acts like WD-40 and helps to lubricate the joints. Synovial fluid contains the gases oxygen, nitrogen, and carbon dioxide. Studies have shown that when you pop or crack a joint, you pull the bones apart, and the capsule surrounding the joint becomes stretched. This causes synovial fluid to squirt from one side of the knuckle to the other. The displaced fluid creates an empty space, which quickly fills with gases. It is this rapid release of gas that causes the popping noise. In order to crack the same knuckle again, you have to wait until the gases return to the synovial fluid, which is why we can't crack the same knuckle twice in quick succession.

Why did Chinese people bind feet?

In China, for almost a thousand years, it was considered to be beautiful for women to have tiny feet. As a result, young girls would have their feet bound to stunt their growth. Foot binding involved tightly wrapping each foot in bandages, sparing only the big toe. The object was to break many of the bones in the foot, to bend the toes into the sole of the foot, and to bring the sole and heel as close together as possible, which sometimes caused toes to fall off. These poor girls would suffer pain for about a year, before their feet eventually became numb. Every few weeks, the bound feet would be squeezed into smaller and smaller shoes. The goal was for the girl to eventually have a foot that measured about 3½ inches (9 cm) long.

Naturally, the girls' feet would become horribly deformed, making walking very difficult. It was usually girls from wealthy families who suffered this ordeal, as poorer girls would have been expected to work, which would have been close to impossible with bound feet. Sadly, this only added to the appeal of the practice, as it meant that tiny feet also indicated that a girl was from a wealthy background. Fortunately, this barbaric custom has been banned for nearly a century.

What causes period cramps?

During her period, a woman's uterus (womb) produces a number of hormone-like chemicals called prostaglandins. These chemicals cause the uterine muscles (the muscles of the uterus) to contract, to push out blood and the endometrial lining. As these muscles contract and tighten, they also constrict the blood vessels that supply the uterus. As a result, blood flow to the uterus is reduced. Because there is less blood flow, this means that less oxygen is brought to the uterus and that waste products such as lactic acid and carbon dioxide are not transported away as quickly. Muscle cramps in general, in any of our muscles, are usually caused by a lack of oxygen, which leads to a painful build-up of lactic acid. In the case of period cramps, reduced blood flow means that less oxygen is brought to the uterine muscles, causing them to cramp.

One way to reduce period cramps is through exercise, probably because exercise releases endorphins, which are chemicals in the body that act as pain relievers and which also make us feel good.

Prostaglandins can also cause some women to suffer from diarrhea and an increase in flatulence (the passing of gas) during their periods. This is because they not only cause the uterus to contract, but they can also affect the intestines. During menstruation, prostaglandins may also cause headaches, backache, and even nausea and vomiting.

Why do our muscles hurt after exercise?

Muscle soreness is generally at its worst during the first two days after exercise and decreases over the following few days. It is thought to be caused by minor tearing of the muscle fibers; as the muscles repair themselves, the soreness disappears. The amount of tearing and soreness depends on how hard and how long you exercise, and what type of exercise you do.

This soreness and stiffness is a normal response to unusual exertion and is part of a process that leads to greater stamina and strength as the muscles recover and build. To help reduce the soreness and stiffness, you should warm up and cool down thoroughly before and after exercise. Gently stretching and massaging the affected muscle will help to increase blood supply to it, and so will aid healing.

What does human flesh taste like?

In 1972, survivors of a Uruguayan plane crash were stranded in the Andes. To avoid starvation, they decided to eat the flesh of fellow passengers killed in the accident. After they were rescued, survivors said that they had cooked the meat briefly and "the slight browning of the flesh gave it an immeasurably better flavor—softer than beef but with much the same taste."

However, infamous murderer Arthur Shawcross, who took the lives of eleven women in New York from 1989 to 1990, said cooked human flesh tastes like nice roast pork.

Whether human flesh tastes like pork or beef, the fact remains that it can be dangerous to partake of such a diet. In the 1960s, there were epidemic levels of a rare and fatal brain disorder called kuru among the Fore (pronounced for-ay) tribespeople who lived in the highlands of Papua New Guinea. Many of them died from kuru during this period, and their deaths are thought to have been caused by the transmission of a virus-like particle, through the tribal practice of cannibalism.

Traditionally, when a member of the group died, he or she would be dissected and wrapped, and then steamed in a fire. During the funeral the brain would be presented to the closest female relative, and she and her children would be given the honor of eating it. Unfortunately, the virus-like particle which causes kuru is found in highest concentration in the brain. Consequently, the Fore's traditional rites were the key factor in the spread of this disease.

The tribespeople believed that kuru was caused by sorcery and could not be convinced that it was due to eating human remains. However, despite this, most of them did stop eating human body parts once they were ordered to do so by police and threatened with imprisonment. Once the cannibalism was stopped, the disease also abated.

Fritz Haarmann (1879–1925) became known as the "Butcher of Hanover" and was thought to have been responsible for the deaths of up to fifty boys and men. After Germany's First World War defeat, Haarmann opened a butcher's shop. The shop prospered, mainly because he sold cheap, fresh meat at a time of great hunger when meat was scarce. After attacking and killing his victims, Haarmann would chop up their bodies and make them into sausage meat, which he cooked and served to his favorite customers. In 1924, a woman who had bought some of his beef became suspicious and contacted the police. The meat was sent to an expert analyst, who somehow concluded that the meat was in fact pork! Nonetheless, the police eventually found grisly evidence for twenty-seven of the murders, and Haarmann was sentenced to death by beheading.

What is the "funny bone"?

The funny bone isn't actually a bone, it's a nerve known as the ulnar nerve. This nerve is connected to the humerus bone, which runs from the shoulder to the elbow, and it's this bone that gives the funny bone its name (humerus—humorous, geddit?). The ulnar nerve runs down the inside part of the elbow. It controls feeling in the fourth and fifth fingers and helps to control the movement of the wrist.

When you bang your elbow against something, you will sometimes knock your ulnar nerve against the humerus bone, causing a sensation of painful tingling and numbness in the arm. There may even be tingling pain felt all the way down to the little finger, where the ulnar nerve ends.

Do adults have more bones than babies?

The skeleton of a newborn baby is made up of more than three hundred parts, most of which are made of cartilage. Over time, most of this cartilage turns into bone, in a process called ossification. As the baby grows, some of its bones fuse together to form bigger bones. By adulthood, the skeleton contains just 206 bones.

Whose bones were found beneath the staircase in the Tower of London?

In 1674, workmen discovered a wooden chest containing the skeletons of two young boys beneath a staircase in the Tower of London. The bones are believed to have belonged to the nephews of King Richard III of England (1452–85).

When he died in 1483, King Edward IV appointed his brother Richard, Duke of Gloucester, as Lord Protector to his twelve-year-old son, Edward, the new king. The young princes, Edward and his nine-year-old brother, Richard, stayed with their uncle Richard in the Tower of London. Richard then imprisoned his nephews and seized the throne for himself, so becoming King Richard III. It is believed that the boys were murdered in the Tower on his orders. Sir Thomas More, who was King Henry VIII's Lord Chancellor some fifty years later, wrote that the Richard III had the boys suffocated with a pillow.

After their discovery, the bones were placed in an urn and taken to Westminster Abbey, where they have remained ever since.

Why do women have larger pelvises than men?

The female pelvis is larger than that of a man because it is designed for childbirth. It is broad and shallow, so that the baby may pass through its bony opening at birth. The pelvis is made up of two hip bones, which create a circular shape and almost meet in the middle (the pubic bone) but don't touch. This gap between the hip bones (pubic area) is connected and held together by the pubic symphysis joint. This joint contains cartilage, which is usually found between bones and allows smooth movement of joints. It is held together by strong ligaments to help ensure these bones remain in place.

During pregnancy, a group of hormones which are collectively called relaxin cause the pubic symphysis to soften and stretch. The joint becomes more flexible, which allows the bones to move freely and to expand to help delivery of the baby during birth.

Why do some skeletons have hair?

Hair consists mostly of a protein called keratin, which is also found in our nails and skin. The word keratin comes from the Greek word *keras,* which means horn, and, in fact, a rhinoceros's horn also consists mostly of keratin. Keratin is insoluble in water, which explains why hair can become blocked in drains for weeks on end. Keratin is a tough substance that resists the enzymes that usually dissolve proteins. Consequently, a corpse's hair will last longer than its flesh, skin, and internal organs.

In 1983, the human remains of a man were found in a peat bog in Cheshire. Scientists found that he had lived during the Iron Age and they named him "Lindow Man" because he was found in a place called Lindow. He was unique because he had a fully preserved beard, unlike any previous "bog body" that had been found. The beard had been trimmed just a few days before his death.

Lindow Man's death had been a violent one. He had been brutally struck three times on the head and then strangled. A small rope had been placed around his neck and tightened, closing off his windpipe and breaking two of the vertebrae in his neck. Scientists also found a gash at his throat, which presumably indicated that it had been cut. It is thought his death may have been either a form of punishment or a ritual sacrifice.

Can muscle turn into fat?

People often talk about muscle turning into fat, but in fact this is about as likely as hair turning into teeth. If a muscle isn't used, it won't convert itself into fat—instead, it will waste away. Anyone who has had a broken leg or arm will almost certainly have seen a clear demonstration of this. After the plaster cast is removed, the leg or arm will appear smaller than before, because the unused muscles have deteriorated. If muscle did turn to fat, the unused limb would be fatter than before and certainly not smaller.

Fat tissue is found directly under our skin. Muscle tissue is located beneath the fat and is attached to our bones—they are completely different tissues. Muscle tissue is made up of muscle cells and contains approximately 70 percent water, whereas fat contains fat cells, which contain less than 25 percent water.

Why do we shiver?

Our bodies work hard to keep our temperature constant at around 99°F (37°C). An abnormally low body temperature, otherwise known as hypothermia, can be dangerous; so if the body senses that we are becoming cold it sends nerve messages from the brain to the muscles, which stimulate them to rhythmically contract and relax—in other words, to shiver. This tightening and loosening of the muscles helps to create

warmth. Our blood carries this warmth around the body, helping to raise our overall body temperature.

Can a head live without a body?

In 1988, the U.S. government granted a patent for a device that
would keep a severed head alive after being surgically removed
from the body. The device has never been used, so it is uncertain how effective it would be. However, the creator of the
device has been contacted by a number of people who want to
know how soon the operation will be available and how much
it will cost. Some of these people are dying or paralyzed, and
many of them say that they would welcome the operation, if
it meant that their minds would remain clear and they could
still think, see, read, remember, talk, and listen.

The proposed procedure would involve attaching the decapitated head to a device essentially consisting of a series of
plastic tubes. These tubes would be connected to the bottom
of the head and neck and would provide oxygen and fluids, as
well as maintaining blood circulation, to keep the head alive.

In 1970, an American brain surgeon called Dr. Robert White
carried out the world's first head transplant, using two mon-
keys. He decapitated both animals and successfully managed
to stitch the head of one monkey onto the body of the other.
The "hybrid" monkey regained consciousness, opened its
eyes, and tried to bite a surgeon who put a finger in its mouth.
It also ate, and it could follow people around the room with
its eyes. However, the monkey was paralyzed from the neck
down because its spinal cord had been severed, and it was im-

possible for the surgeons to reconnect the numerous nerves necessary for it to regain any bodily movement. The monkey survived for about seven days after the transplant.

White claimed that this surgery could benefit paraplegics, who may die as a result of the long-term medical complications that often accompany extensive paralysis. He believed that if these people were to receive new bodies, donated by patients who were brain dead but otherwise physically healthy, it would give them a new chance of life, even though they would remain paraplegic.

Is it possible to shrink heads?

In the year 1599, around 25,000 Spaniards were brutally murdered by the Jivaro tribe, who lived in Ecuador and Peru. The heads of the victims were put through a special process to shrink them and then were kept as trophies.

The explorer F. W. Up de Graff described the following event, which took place centuries later, in 1897. The Jivaro tribe raided another tribal settlement, and Up de Graff and some other Europeans followed, but didn't take part in the raid. The Jivaro killed some of the members of the tribe, scared off the rest, and then set about chopping off the heads of their victims. One victim was a woman who had the misfortune of not being dead. Untroubled by this, one raider held her down, another pulled back her head, and a third tried to cut off her head with a stone ax. It took a while to cut through her neck, so the Jivaro asked to borrow Up de Graff's machete. The explorer handed it over, rationalizing that it would put the victim out of her misery.

The Jivaro carried the severed heads back to camp, holding them either by the hair or by a strip of bark passed through the mouth and out through the neck. They carefully made a slit up the back of the head from neck to crown, peeled off the skin, then turned it inside out and scraped it with a knife. The skull was thrown away and the skin, scalp, and hair were cooked in pots filled with boiling water, which made them shrink. The slit at the back of the scalp, the eyelids, and the mouth were either sewn up or plugged. Heated pebbles or hot sand were introduced through the neck, and the head was shaken until the skin became like leather. The head was then tanned and stuffed, and although it resembled its victim, it was only the size of a large orange. The hairs of the face, eyebrows, and eyelashes were now disproportionately long, so they were singed off. The shrunken heads were shown off to fellow tribal members when festivals were held.

There are even accounts of head-shrinking from the twentieth century. During the Second World War, the Nazis displayed two shrunken heads in the Buchenwald concentration camp, in order to intimidate the prisoners. In 1942, Nazi high command ordered staff to stop shrinking heads unless it was for "medical reasons." This order wasn't out of any feelings of guilt, but rather out of concern that word of their genocide might leak out, as visitors were taking away shrunken heads as souvenirs. One of the shrunken heads was later used as evidence of the Nazi atrocities during the Nuremburg trials.

Why do some tribal women wear brass rings around their necks?

Women of the Karen tribe in Thailand wear many brass rings around their necks, a practice which began originally to help protect their necks from predatory tigers (nowadays, of course, it is the tigers who are endangered). Over time, it became a sign of beauty for women to have long necks and to wear a foot-high stack of metal that covered the whole neck.

The process begins when a young girl is fitted with a heavy brass or iron neckband on her fifth birthday, and every few months more bands are added, until the neck is 12 inches (30 cm) long. In adulthood, some of these women carry over 28 pounds (13 kg) of metal around their necks.

Contrary to appearances, the metal rings do not in fact cause the neck to grow longer; rather, they push down on the shoulders, making the neck appear more prominent. After the women have worn these bands for a while, their neck muscles become unable to function—without the bands, the women would be unable to hold their heads up straight. If a girl had her first band fitted at a young age, meaning her neck muscles had never had the chance to properly develop, she might even be at risk of suffocation if they were ever removed.

Is it true that X-rays were once carried out in shoe shops?

It is ironic that X-rays can both cause cancer and be used to treat it. Nowadays, with the use of very small doses of radiation to produce high-quality X-ray images, the risk of cancer after properly supervised X-ray examinations is extremely small.

Between the 1930s and 1950s, a device called the shoe-fitting fluoroscope was a common fixture in shoe shops. It was a unit that usually consisted of a vertical wooden cabinet with an opening near the bottom into which the feet were placed. When you looked through the viewing holes on the top of the cabinet you would see a fluorescent image of the bones of the feet and the outline of the shoes. When the feet were in the shoe-fitting fluoroscope, the customer was effectively standing on top of an X-ray tube. The fluoroscope helped to measure shoe size and tested the fit of a new pair of shoes. When it was realized that X-rays could be harmful, the use of the fluoroscope declined.

The fluoroscope was widely used by doctors to view inside the body. X-rays were also used to shrink infected tonsils, and thousands of children received this treatment. Decades later, it was realized that this treatment could cause thyroid cancer.

What was the condition suffered by the Elephant Man?

Joseph Carey Merrick (1862–90), also known as John Merrick, was better known as the "Elephant Man." In early childhood he developed a number of grotesque deformities, and in later life he made a living by exhibiting himself at fairs. He was discovered by Dr. Frederick Treves and became a patient at the Royal London Hospital. He remained there until 1890, when at the age of twenty-seven he died in his sleep of accidental suffocation.

Merrick's preserved skeleton is still in the Royal London Hospital but is not on public display. In 1987, it was reported that singer Michael Jackson had offered the hospital a million dollars for the skeleton, but they refused to sell.

Merrick suffered with a condition that caused him to be horribly disfigured. He had huge folds of warty skin that dangled from his chest, and parts of his body were abnormally large, including his head. His whole body was affected, apart from his left arm and his genitals. It is believed that he suffered from a very rare and severe case of a condition called Proteus Syndrome. Proteus was a Greek god who could change his body into different shapes to escape from his enemies. Proteus Syndrome causes the overgrowth and enlargement of the hands, feet, and skull. Other symptoms include numerous benign tumors and raised, rough skin. It is such a rare condition that since Dr. Michael Cohen identified it in 1979, there have only been around two hundred reported cases. Even so, Merrick had an unusually severe case. There is no specific medical treatment for the syndrome, but nowadays abnormal growths can be removed surgically. Researchers are still trying to find the cause of this condition, but as yet there is no cure.

What was unusual about King Charles VIII of France's toes?

Toward the end of the fifteenth century, King Charles VIII of France is said to have made it fashionable for men to wear shoes with square toes that looked rather like a duck's beak. He himself had six toes on both feet, and for this reason he decreed that very wide shoes were something to be admired. These shoes were known as "bear paws," "crackowes," or "duckbill shoes," and over time they became wider and wider, until they reached 12 inches (30 cm) across the ball of the foot. Men who wore duckbill shoes tended to adopt a waddling gait because the shoes were so wide.

Do men have one rib fewer than women?

Men and women both have twelve pairs of ribs, whose purpose is to protect our internal organs, such as the heart and lungs. The belief that men have one less rib than women probably comes from the account in the Book of Genesis (2:21,22), "And the Lord God caused a deep sleep to fall upon Adam, and he slept; and he took one of his ribs, and closed up the flesh instead thereof. And the rib which the Lord God had taken from man, made he a woman."

Can a human being spontaneously combust?

Human spontaneous combustion is a mysterious, controversial, and much derided phenomenon, in which a person is said to suddenly burst into flames, without the presence of any external fire or heat. Cases have been reported of burned corpses being found, their bodies charred, but the furniture around the victim seemingly untouched by the fire. This phenomenon has remained a contentious mystery for hundreds of years, with a number of possible explanations being suggested, including balls of lightning or a buildup of methane inside the intestines.

However, scientists now believe that they may have found the answer, in a theory known as the "wick effect." The theory is that in certain rare and particular circumstances, a human body can burn in a way comparable to a candle. A source such as a lit cigarette may start the fire, and it is believed that body fat can act as fuel to keep the body burning. A group of researchers carried out an experiment to demonstrate the wick effect. It involved setting light to a dead pig wrapped in cloth, which was designed to represent a person wearing clothes. The pig burned for many hours, and the charred effect was similar to that found in apparent cases of spontaneous human combustion. The scientists believe they demonstrated how a case of spontaneous human combustion could occur to a person who had already been knocked unconscious. It could also explain why only part of the body—the part that is rich in fat—burns, while the rest stays intact.

Have human beings ever had a tail?

Many evolutionists believe that distant ancestors of humans possessed a tail. Indeed, between four to seven weeks of development, humans do indeed have a tail, which is later reabsorbed into the body. An adult has a coccyx, or tail bone, which is usually made up of four small bones, which are fused together to create one bone at the lower end of our vertebral column (spine).

In 1901, Ross Granville Harrison, M.D., who was an associate professor of anatomy at an American university, described a healthy infant whose tail grew very quickly. At birth it was 1½ inches (4 cm) long, and it grew to nearly 3 inches (8 cm) by the time the child was three months old. If the boy coughed, sneezed, or was irritated, the tail would move. The tail was eventually removed, and when Dr. Harrison examined it, he found it was covered in normal skin and contained fat, nerves, and blood vessels. It also contained strands of muscle, which explains how the tail could move.

A man called Chandre Oram is revered in some parts of India for having a 13-inch (33 cm) tail. He has a large following as many people believe, because of his tail, that he is an incarnation of the Hindu monkey god Hanuman. People have even reported being cured of ailments after touching his tail. Doctors have offered to remove Oram's tail surgically, but he has declined their help. They believe his tail is due to a congenital condition called spina bifida, which results from the incorrect development of the spinal cord in the womb.

CHAPTER FOUR

Hematological Hors D'oeuvres and Aortic Amuses-bouches

What is bloodletting?

Bloodletting is a medical procedure that used to be thought to help cure disease, and sometimes involved draining large quantities of blood from the body. For around two thousand years, from the first century B.C. until the mid-1800s, bloodletting was the main therapy used by doctors. Practitioners would cut their patients, using items such as scissors and scalpels, and then collect the resulting blood in a bowl. They would continue to drain blood until the patient became faint.

In the Middle Ages, bloodletting would often be carried out by the local barber. This is why the traditional barber's pole is colored red and white—the red color signifies the blood and the white represents the bandages. Members of the Barber Surgeons' Guild were licensed to work with blades, which meant they were qualified to do anything from shaving off your mustache to cutting off your leg. A barber surgeon's examination of a patient would usually involve taking a urine sample. He would make his prognosis by using a chart known as a urine wheel, which showed what the various shades of urine could indicate about the patient's health. The prognosis was not good if your urine happened to be a dark color, as this apparently signified death! The barber would also often smell and taste the urine. The treatments offered included diet and exercise advice, laxatives, induced vomiting, diuretics, and bloodletting.

Perhaps unsurprisingly, it wasn't unusual for patients to die after undergoing bloodletting treatment. George Washington (1732–99) had nine pints of blood drained in twenty-four hours to help with his throat infection, but died soon afterward. The poet Lord Byron died in 1824 because his doctors literally bled him to death.

How can leeches be used to help severed limbs?

From the Middle Ages to the present day, leeches have been used to treat a variety of medical conditions. In the Middle Ages, the popularity of leeches meant that some people had jobs that involved collecting them to sell to doctors. Leeches used to live in marshy areas all over Britain, including the Lake District and the Somerset Levels. The job of a leech collector involved wading through water, bare-legged among the reeds, and waiting for the leeches to latch themselves onto his legs. A bite from a leech is a little uncomfortable and leaves behind a mark that is similar in shape to the Mercedes-Benz logo. Around 5 fluid ounces (150 ml) of blood can be lost from each leech bite, so leech collectors would probably have suffered from dizziness, and a lot of bleeding from the wounds. Also, leeches can infect their host with the bacterium *Aeromonas hydrophila*, which causes diarrhea and infection.

In the 1800s, leeches were used to help treat a wide range of conditions including colds and other infections. Large numbers of leeches would be applied to areas of inflammation or pain. Women often had leeches placed in their vaginas, as this treatment was thought to help with conditions as diverse as vaginal discharge and cervical cancer. Leeches were also applied to the clitoris to treat a condition known as "exquisite sensibility of the clitoris," and this became a common treatment for many female complaints.

Also in the 1800s, the application of leeches to treat disease became so popular that leeches became an endangered species in Europe, and the word "leech" became a slang term for doctors themselves. After the discovery of germs such as bacteria, the use of leeches declined, although they are now making a comeback.

Leeches are often used today as a tool for healing skin grafts or restoring circulation. The leech's saliva contains substances that anesthetize the wound and widen the blood vessels, which increases blood flow and prevents the blood from clotting. During operations such as reattaching a body part, surgeons find it difficult to stitch up veins, as their walls are very thin and they are often badly damaged. When trying to reattach a finger, for example, if the blood doesn't flow freely it can become congested or stagnant, which can lead to the finger's becoming blue and lifeless, until eventually the tissue dies. In cases like this, doctors use leeches to stimulate a flow of blood through tissue where there is a congestion of blood. The stimulation of blood helps to clear the congestion. Treatment with a single leech can last for hours, maintaining blood flow in the tissue for all of that time.

Did Dracula really exist?

In 1897, the novelist Bram Stoker (1847–1912) published his famous vampire novel, *Dracula*, whose main character was of course fictional. In the novel, Dracula bites the necks of his victims to drink their blood. However, it is believed that Dracula may have been based on a real-life person, an evil man who lived in the 1400s, who himself went by the name of Dracula.

In medieval times, central Europe consisted of a large number of small, autonomous states, one of which was Wallachia, in what is now Romania. In the 1400s, Wallachia was ruled by Vlad III , who liked to be known as Dracula. Dracula was keen to keep his country free of crime, enemies, and just about anyone he considered not to be an ideal citizen within his kingdom.

When he became ruler, he invited the poorest people and those with diseases and disabilities to a large banquet and told them they would be free of hunger, worries, and aches and pains in the future. They were given food and drink, and afterward he left the hall. On Dracula's order, the soldiers then barricaded the doors and set fire to the building, killing all the guests.

This Dracula did not suck blood from his victims, but he did take great pleasure in torturing people and would impale his enemies on long, wooden spears, which resulted in a long, agonizing death (hence his other nickname, "Vlad the Impaler"). He impaled up to 100,000 people, and enjoyed arranging them into patterns so they looked "arty." Sometimes, for his own amusement, he would skin them alive or hack them into little pieces.

Dracula died while fighting in 1476, and his head was stuck on a pole and displayed for all to see. The rest of his body was buried at a monastery, but when archaeologists tried to dig it up in the 1930s, the coffin was empty!

Would the heart keep beating if it were separated from the body?

In the 1984 film *Indiana Jones and the Temple of Doom*, the evil cult leader Mola Ram rips out his victim's still-beating heart. The heart continues to beat, even though separated from the body. In real life, it isn't quite as simple to remove the heart, as the ribcage would need to be broken in order to access it. However, the heart would continue to move if removed from the body, although to describe it as "beating" might be slightly inaccurate. If a beating heart is removed from a human body, it will continue to beat for a few seconds, and after this it will fibrillate for about three to five minutes. "Fibrillating" means that the heart continues to make irregular, rapid twitching movements.

Is it true that people used moss taken from a dead person's skull to help cure nosebleeds?

In the 1600s, chemists in England sold a gray-green colored moss called usnea to help patients who suffered with conditions such as nosebleeds and nervous disorders. In the first edition of the *Pharmacopoeia Londoniensis* in 1618, under the direction of Sir Theodore Mayerne (1573–1655), usnea was described as a drug made from the moss that grew on the skull of man who had died a violent death (such as criminals who had been hanged). It was preferred that the corpse had been left to rot for a long while because this made it easier to produce the remedy from the moss. Usnea was also recommended in the 1650 revision of the *Pharmacopoeia*.

Usnea was reportedly used over three thousand years ago

in ancient Egypt, Greece, and China to treat infections. Other ancient remedies include:

- Urinating in an open grave to cure incontinence.

- Sticking a twig of an elder tree in the ear, and wearing it night and day, to cure deafness.

- Tying a hairy caterpillar in a bag around a child's neck, to cure whooping cough.

- Binding the temples with a rope with which a man has been hanged, to relieve a headache.

However, unlike the bizarre treatments listed above, usnea may have actually had some medicinal value as it contains usnic acid, which is a powerful antibiotic and antifungal agent.

Do bedbugs really bite?

Sleep tight, don't let the bedbugs bite! Unfortunately, bedbugs do bite us, but they rarely cause any harm. Bedbugs are wingless insects, roughly oval in shape. They grow up to about ¼ inch (5 mm) long, and they are fast runners. They like to live near where people sleep, which includes mattresses, bed frames, carpets, floorboards, and any other crack or crevice they can find.

Bedbugs mostly bite during the night. Their mouthparts are especially adapted for piercing skin and sucking blood, but we don't usually notice because their bite is subtle and painless. They inject saliva during feeding, which contains anticoagulants, so that the blood doesn't clot and the food keeps on coming. They also inject anti-inflammatory agents, so you are less likely to feel irritation or get a pimple, which could cause you to wake up and scratch and interrupt their meal! After eating a meal of blood, bedbugs change from a rust-brown color to a deeper red-brown.

Is it true that a heart transplant can cause a change in personality?

Medical opinion is divided about whether or not organ recipients can take on the personality traits of their donors. Although many experts are skeptical about this phenomenon, there is anecdotal evidence that suggests that recipients of organ donations might actually take on some characteristics of the donor.

In her forties, American Claire Sylvia suffered from an incurable and progressive heart disease, which meant she was housebound and terminally ill. Her only hope was a heart and lung transplant, and the donor was an eighteen-year-old male motorcyclist who had died from severe head injuries.

Fortunately, Claire survived the operation, but noticed

certain changes in her personality. She found she now had cravings for beer and chicken nuggets, which she had previously disliked. She also had strange and vivid dreams about a young man she didn't recognize. Although she had been heterosexual prior to the transplant, she became attracted to women, especially if they were blonde. She met with the donor's family, and it was confirmed that her new personality traits matched those of the donor and that he was the man who had appeared in her dreams.

Prior to receiving a heart transplant, William Sheridan, a retired catering manager, had had the artistic ability of a young child. After his transplant, he found he was suddenly blessed with artistic talent, producing beautiful drawings of wildlife and landscapes. He later learned that the man who had donated his heart was a keen artist. When asked if her son was artistic, the donor's mother told him, "He was very artistic. He showed an interest in art when he was just eighteen months old. He always preferred to be given art supplies rather than toys."

Why do professional runners train at high altitudes?

At very high altitudes there are fewer oxygen molecules in the air than at sea level. As a result, a climber at high altitude will breathe faster to try to make up for the lower oxygen levels. However, after the climber has been at a high altitude for a few days, his or her body begins to find other ways to help get more oxygen to its cells. One way involves the body's producing more red blood cells. Climbers who have ascended Mount Everest have been found to have 66 percent more red blood cells than normal. If the blood has more red blood cells, it can carry more oxygen around the body.

Many professional runners train at high altitude to increase the number of their red blood cells. With more red blood cells there is more oxygen transported around the body, so the runners' performances improve when they race at sea level. After two to three weeks at sea level, the number of red blood cells returns to normal levels.

Can you catch a cold by standing out in the rain?

Contrary to the old wives' tale, sitting in a draft does not cause a cold—nor does cold weather, wet hair, or standing in the rain. Colds are caused by cold viruses, of which there are around two hundred. Studies have been carried out that show that people are no more likely to catch a cold as a result of cold or wet conditions. One particular study involved people being sent out for walks in the rain and then returning to an unheated room. They were not even allowed to dry themselves with towels, so they remained in the room cold and wet. Despite these conditions, the group showed no increase in their susceptibility to colds.

If coldness and wet conditions were responsible for causing a cold, we would expect Eskimos to suffer from them permanently. In fact, they do not. The Arctic and Antarctic are relatively germ-free, and explorers have reported being free from colds until coming into contact with infected people.

The common cold is caused by viruses that can be passed from person to person. They can be caught through droplets in the air after someone has sneezed, but more commonly through touching an object such as a door handle. For instance, after a person with a cold has rubbed his or her nose he or she might touch the door handle. Another person might later touch the same door handle, and then rub his/her nose. Bingo! The virus has been passed, and a cold may follow.

The reason we catch more colds in winter is not because of the weather, at least not directly. Rather, it is because we tend to stay indoors more, which means we spend more time in close proximity to other people, and we tend to keep our windows closed, trapping any viruses in the room.

Will sitting on a cold surface give you hemorrhoids?

Another old wives' tale is debunked: there is no evidence to suggest that you will get hemorrhoids from sitting on a cold surface. Hemorrhoids, also known as piles, can be extremely uncomfortable, and more than half the U.K. population will suffer from them at some time in their lives. Hemorrhoids are enlarged, swollen blood vessels in or around the lower rectum and anus. They sometimes cause bleeding, which may be noticeable on toilet paper after wiping the bottom.

The myth that hemorrhoids are caused by sitting on a cold surface was probably designed as a way to get lazy people to get up and do some work. However, prolonged sitting and straining on the toilet can lead to the development of hemorrhoids or make them worse. To avoid hemorrhoids, you should eat plenty of fiber, which eases the passage of waste matter through the large intestine and anus and so helps prevent their development.

If creams or suppositories do not clear up the hemorrhoids, then "banding" may be recommended. Banding is a procedure that is carried out by a surgeon and involves placing a tight rubber band around the base of the hemorrhoid, which cuts off its blood supply. After a few days, the hemorrhoid dies and falls off.

How mad was Mad King George?

King George III (1738–1820) ruled England in the eighteenth century, and was also known as Mad King George. He is thought to have suffered with porphyria, which is a group of rare blood disorders in which substances called porphyrins build up in the blood. This condition can cause skin rashes, abdominal pain, and mental confusion.

King George is largely remembered for the periods when he lost his mind. He would sometimes become so violent that he would have to be placed in a straitjacket and chained to a chair. In one moment of madness, he pulled off his wig and ran around naked and feverish. Another incident occurred when he was being driven through Windsor Park and told the driver of the carriage to stop. The king got out and walked over to an oak tree, shook hands with a branch, and talked to it for several minutes. He thought he was talking to the King of Prussia. Despite his illness, George lived on into his eighties, but by the time he died he was blind, deaf, and mad.

There is recent evidence that suggests that King George's madness may have been caused by poisoning. Some strands of his hair were recently tested and found to contain a great deal of arsenic—over three hundred times the toxic level. Researchers also found that, according to the king's medical records, the attacks didn't occur until he was fifty years old. Porphyria attacks can be triggered by many substances, including arsenic. In his medical records, there is a reference to arsenic being used as a skin cream and wig powder, and he was also taking a medicine that contained arsenic to control his madness. Ironically, it seems that this medicine may actually have been exacerbating George's condition.

What would happen if we were given the wrong blood type during a blood transfusion?

In 1667, Jean Denys, who was Louis XIV's physician, carried out the first human blood transfusion. He transfused 9 fluid ounces (255 ml) of blood from a lamb into a fifteen-year-old boy who was suffering with a fever. Before the transfusion, "[the boy's] wit seemed wholly sunk, his memory perfectly lost and his body so heavy and drowsy that he was not fit for anything," but soon afterward he displayed "a clear and smiling countenance." The improvement in the boy's condition was probably due to the fact that his body was winning the fight against the infection rather than as a consequence of receiving sheep's blood. Six months later, Denys carried out another blood transfusion, which involved giving his patient the blood of a calf. Unfortunately, this attempt was less successful, and the patient died after the third treatment.

A blood type (also called a blood group) is a classification of blood based on the presence or absence of certain molecules on the surface of red blood cells. This classification is known as the ABO system and was discovered in 1901 by Karl Landsteiner while he was trying to understand why blood transfusions sometimes saved lives and at other times caused death. There are four main blood groups, which are called A, B, AB, and O. These groups can be either positive or negative, meaning that a person's blood group can be one of eight variations.

In the U.K., O is the most common blood type. Type O is known as the universal donor, because O-type blood can be given to any person with type A, B, AB, or O blood. However, if a patient has blood type O, he or she can receive only type O. Blood type AB is called the universal recipient, because a patient with this type can receive blood from all blood groups. The blood group you belong to depends on what you have inherited from your parents.

Blood plasma is the liquid part of the blood, which consists mainly of water. Many substances travel in the blood plasma, such as red blood cells, which transport oxygen around the body. The differences between different blood types are defined by the presence or absence of certain protein molecules, called antigens and antibodies. Antigens are found on the surface of red blood cells, and antibodies are found in blood plasma. Different people have different types and combinations of these molecules.

During a blood transfusion, if the blood group of the donor and patient (recipient) are not compatible, the recipient's immune system will attack the donor blood, causing the red blood cells from the donated blood to clump, or agglutinate, which results in the formation of blood clots. These blood clots can block blood vessels and stop the circulation of the blood to various parts of the body, which can have fatal consequences for the patient.

How do vaccinations work?

Vaccines, which are also known as immunizations, help to prevent illness or death for millions of people each year. A doctor called Edward Jenner (1749–1823) was the first person to produce a vaccine when he developed the vaccine against smallpox. He noticed that milkmaids who had contracted and recovered from a mild disease called cowpox, which caused blisters on their hands, never caught the deadly disease smallpox. He persuaded a mother to let him inject her eight-year-old son with cowpox. The boy caught the disease and then recovered. Jenner next injected the boy with smallpox and found that the boy was protected against the disease because he had previously been exposed to cowpox. The cowpox virus is so similar to the smallpox virus that our bodies cannot tell the difference. Jenner realized that by injecting people with harmless cowpox, he could protect them from the deadly smallpox.

Vaccines prepare the body's immune system to fight disease by taking advantage of the fact that the immune system can "remember" infectious germs. Vaccines, which are often injected into the body, give us immunity without our having to experience the disease or its symptoms. Each vaccine contains a dead or weakened form of the germ (usually a virus or bacterium) that causes a particular disease. For example, there are two vaccines against typhoid fever: one vaccine contains dead *Salmonella typhi* bacteria, and the other contains a live but weakened strain of the *Salmonella* bacteria that causes typhoid fever. Even though the germ in the vaccine has been altered so that it won't make you ill, the part of the germ that stimulates your immune system to respond (the antigen) is still present.

After you have been vaccinated, some of the white blood cells that are responsible for protecting you against disease, called B lymphocytes, detect the antigens in the vaccine. The B lymphocytes respond as if the real infectious germ has invaded the body. They multiply to form an army and develop into either plasma cells or memory B cells. The plasma cells produce antibodies, which attach to and inactivate the bacterium you are being vaccinated against. Over time, the antibodies will gradually disappear, but the memory B cells will remain in your body for many years. If the body comes into contact with the same bacterium in the future, the antibodies and memory B cells will recognize and fight it—therefore, the person will not become ill.

Why are some people allergic to house dust mites?

Most of the dust found in a house is made up of dead skin cells, which are continually being shed from the skin into the air.

The house dust mite lives in dust and eats dead skin cells and hair. These microscopic bugs have eight hairy legs, and when magnified they look like something out of a horror film.

Dead skin cells and dust mites are mainly found in mattresses, furniture, and carpets. A mattress may contain up to ten million dust mites, and many thousands can be found on a single pillow. Nearly a hundred thousand mites can live in 1 square yard (0.8 sq. m) of carpet. A single dust mite produces around twenty poos each day, each of which contains a protein that can cause an allergic reaction, such as itchy eyes, eczema, and even asthma attacks.

Why do ticks become so attached to us?

Ticks are small, blood-sucking, eight-legged parasites, which are often found in tall grass, where they wait patiently on the tip of a blade, ready to attach themselves to a passing animal or human. The tick uses its mouthparts to burrow into the skin of its host, so that it can feed on its blood. Ticks' mouthparts consist of hooks, which make it very difficult to remove a tick from your skin. The problem with ticks is that their bite may pass on dangerous diseases such as Lyme disease, which can develop into a form of arthritis but is treatable with antibiotics.

Can we live without a spleen?

The adult spleen is found on the left side of the abdomen, behind the stomach. It is oval in shape and is about the size of a large apple. The spleen acts as a reservoir for blood, so that if blood is needed elsewhere in the body, perhaps because of a hemorrhage, it can be quickly diverted there from the spleen. Special white blood cells are produced in the spleen to fight infection, so if the spleen were removed, the body's resistance to infection could be lowered.

The spleen can be easily damaged by trauma such as injury. If this happens, a hemorrhage can occur, so the damaged spleen has to be removed quickly. However, the body can survive without it.

Can a dog lick help to heal a wound?

When a mother dog licks her newborn puppies, she is helping to protect them against disease, as the saliva in her mouth contains antibacterial properties, which help to fight certain bacteria that could cause illness. She also licks her own nipples, and this too is designed to keep the puppies free from disease.

There is evidence to suggest that a dog's saliva can prevent the growth of certain bacteria such as *Escherichia coli* (E. coli), which causes severe cramps and diarrhea in humans and can be transmitted by infected food or drink. However, dog's saliva is not an effective antiseptic, and it should not be used for healing human wounds because it contains many types of bacteria that could be harmful.

In the 1950s, research showed that a dog's saliva contained a protein called epidermal growth factor (EGF), which helps to promote more rapid healing. When animals lick their wounds they are constantly applying EGF, resulting in a shorter healing time. But when doctors applied EGF directly to human wounds, they found it wasn't effective, because our bodies contain enzymes that destroy these growth factors in a matter of minutes.

How does regular exercise affect the heart?

When a person is at rest, his or her pulse rate (otherwise known as the "resting heart rate") ranges between about 60 and 80 beats per minute. If a person exercises regularly, it can lead to several body changes including a lower pulse rate. As the heart muscle, along with other muscles in the body, increases in size and becomes more powerful and efficient, one beat of the heart is able to push more blood per minute. During exercise, the blood pushed out of the heart of a trained person is around 6½ gallons (30 liters) per minute, whereas an untrained person's heart can push around only 5 gallons (21.5 liters) per minute. As a result, the trained person will have a much lower pulse rate than the untrained person. The trained person's heart is able to pump more blood per minute, as it is stronger and more efficient than the untrained person's heart. It has been shown that the heart can increase in size by as much as 40 percent through exercise. Retired cyclist Lance Armstrong, who won the Tour de France a record seven times, had a resting heart rate of 32 to 34 beats per minute while in training.

If all the blood vessels in one person's body were laid end to end, how far would they reach?

Blood vessels are tubes that carry blood around the body and include arteries, veins, and capillaries. It has been estimated that there are over 60,000 miles (97,000 km) of blood vessels in a child's body, and around 100,000 miles (161,000 km) in an adult's. The circumference of the earth at the equator is 25,000 miles (40,000 km). Therefore, if the blood vessels of an adult were lined up end to end, they would circle the earth four times!

Most of the blood vessels in the human body are microscopic capillaries, which are about a hundredth of a millimeter (a tiny fraction of an inch) thick. A drop of blood contains millions of red blood cells, and the capillaries are so small that the red blood cells can only pass through them one by one. Around 40 billion capillaries (that's 40,000,000,000!) make up most of the body's 100,000 miles (161,000 km) of vessels.

Can a large waist size increase the risk of heart disease?

Studies have shown that people with the largest waist sizes have the highest risk of developing type 2 diabetes and heart disease, and in the case of type 2 diabetes being overweight is the main risk factor.

Type 2 diabetes is a disease in which the body's cells no longer respond to insulin. (Insulin is a substance released by the pancreas, which regulates sugar levels in the blood, but because the cells of a type 2 diabetic no longer respond to insulin, the body fails to control the amount of sugar in the blood, which can lead to serious health problems.) It is believed that fat cells that develop around the waist pump out chemicals that damage the insulin system, and the more fat cells a person has, the more the body's cells become resistant to the body's own insulin. Losing weight, especially around the waist, makes cells more receptive to insulin, and so helps the body to bring blood sugar levels within a normal range.

Women with waists greater than 35 inches (89 cm) and men with waists greater than 40 inches (102 cm) are classed as being at a higher risk of contracting heart disease and diabetes than people with a smaller waist size. Scientists believe that people's waist measurements can predict more accurately their risk of contracting type 2 diabetes and heart disease than their weight. It has been suggested that even a smaller waist size of 32 inches (81 cm) in women or 37 inches (94 cm) in men may significantly raise the risk of these diseases.

In 2001, a study was carried out on 9,913 people between the ages of eighteen and seventy-four, which concluded that, for maximum health a man needs to keep his waist size at no more than 35 inches (89 cm). Men with the biggest bellies were at the greatest risk.

Alongside its neighbor Nauru, Tonga has the fattest population in the world. In Tonga, 92 percent of all over-thirties are overweight or obese, and this is affecting the nation's health. Almost 20 percent of the adult Tongan population suffers from diabetes, and the death rate from diet-related illnesses is ten times that of the U.K. This is mainly due to the Tongans' diet, as they eat large amounts of saturated fats. Many Tongans like to eat corned beef, and their version of corned beef contains double the fat of the type typically eaten in the U.K. Another staple is a dish called the "lamb flap," made up of lamb belly, which is up to 50 percent fat. On special occasions, such as church conferences, suckling pig is served, and people gather for three feasts a day in celebrations that go on for weeks on end.

However, in Tonga there isn't the same kind of stigma attached to being fat as there is in the U.K., and many Tongans find large bodies very attractive.

Can you catch an infection from using a public telephone?

When a person with a cold coughs or even talks, tiny droplets of moisture fly out of his or her mouth, carrying the cold virus into the air. If someone suffering with a cold talks into a telephone, viruses will land on the mouthpiece and sit there. However, unless the next caller actually touches the mouthpiece with his or her lips, there is no way for the virus to get into his or her body.

Nonetheless, telephones probably do pass on disease. But these illnesses are transmitted by the handset of the telephone, not the mouthpiece. When a sick person coughs into his or her hands or rubs his or her nose, the virus or bacterium lands on his or her hand, and can then be passed to the telephone handset when a phone call is made. When the next person makes a call, he or she may pick up the tiny bacteria or virus particles on his or her hands. Later, if this person touches his or her eyes, nose, or mouth the germ can enter his or her body and infection can result. Simply washing your hands can eliminate most of the colds and other illnesses you ever come into contact with.

Why do we get fever?

When we suffer with an infection, such as a cold, we may also experience a fever. Fevers are caused by chemicals flowing in the bloodstream that make their way to a region of the brain called the hypothalamus, which helps to control the body temperature. As a result of these chemicals entering the hypothalamus, the body temperature then rises.

One purpose of a fever is thought to be to raise the body's temperature enough to kill off certain bacteria and viruses that are sensitive to temperature change. However, the fever itself can be deadly in some cases, so we tend to use medicines such as aspirin to reduce it.

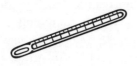

Who ate King Louis XIV's heart?

In the 1800s, an English eccentric called Frank Buckland worked as a surgeon and naturalist. On his estate, Buckland kept an edible menagerie and often cooked himself bizarre dishes such as mice on toast, elephant-trunk soup, rhinoceros pie, and slug soup. Buckland also ate grilled porpoise head, but he confessed he "could not stop it tasting like an oil-lamp wick." Some of his worst dishes included earwigs, stewed mole, and bluebottles. It's even said that when he found out that a leopard at the London Zoo had died, he rushed to where it had been buried and dug it up, so that he could cook its meat.

Buckland also kept a collection of relics, including a lock of hair from Henry IV. His most precious object was the embalmed heart of King Louis XIV. During a meal with a friend he said, "I have eaten many strange things in my lifetime but never before have I eaten the heart of a king," and with that, Buckland began eating away at the king's heart.

Which worm likes to drink human blood?

The hookworm is a parasitic worm about ½ inch (1 cm) long and creamy white in color. It lives in the small intestine of its host, which may be a mammal such as a dog, a cat, or a person. Two species of hookworm commonly infect humans: *Ancylostoma duodenale* and *Necator americanus*. *Ancylostoma duodenale* is native to parts of North Africa, northern India, and parts of western South America. *Necator americanus* is found in central and southern Africa, southern Asia, Australia, and the Pacific Islands. Between them, they are thought to infect around 800 million people worldwide.

Lying on a beautiful beach in a tropical country is most people's idea of heaven, but it is also a great place to pick up a hookworm, as they love warm, damp sand. When we sit or lie on sand, the larvae of the hookworm, which are barely visible, can bore through the skin of our feet or thighs. Once in our system, they are carried in the blood and eventually reach the small intestine, where they take a big bite of the intestine wall, and then hang on. They bite into blood vessels so they can eat their favorite food—blood. The adult worms produce thousands of eggs, which are passed out in our poo. If we poo onto soil or into water, these eggs can develop into larvae and then infect someone else.

The first sign of hookworm infection is usually an itch or rash at the site where the larvae penetrated the skin. A light infection might create no symptoms, but heavy infection can lead to anemia, abdominal pain, and diarrhea. In countries where hookworm is common and reinfection is likely, light infections are often not treated. Nonetheless, there are effective antiparasitic drugs that can kill the hookworm.

CHAPTER FIVE

Scatological Silliness

Why doesn't sweet corn get digested properly?

If you notice a yellow-dotted poo in the toilet, there is a good chance that the poo contains sweet corn. The outer coating of a sweet corn kernel cannot be digested, because our bodies don't have the enzyme necessary to break it down. However, the inside of the sweet corn kernel, which consists mainly of starch, is easily digested by the body. Undigested foods, such as the outer part of sweet corn, become mixed up with the rest of the poo, which is mainly water, dead cells, and bacteria. Similarly, eating pimientos will produce red blotches in the stools, as they pass through the digestive tract almost unchanged.

What did people use to wipe their bums before toilet paper was invented?

Toilet paper was first used in China in the late 1300s. The sheets were enormous, measuring 2 feet by 3 feet (0.6 m by 1 m), and were mainly used by emperors. Modern toilet paper was invented by an American called Joseph Gayetty, who first packaged bathroom tissue in 1857. To ensure that he wouldn't be forgotten, he had his name printed on every sheet. It was called "therapeutic paper" because it contained plant extracts of aloe. In 1879, a British man called Walter Alcock invented perforated toilet tissue on a roll. Before paper was used to wipe the bottom, people had lots of other innovative ideas:

- Wealthy Romans used a sponge tied to the end of a stick which had been soaked in salt water to clean their bums. It is thought that this is where the expression "the wrong end of the stick" has come from.

- The Japanese used wooden sticks called *chu-gi* to clean their bums.

- Early Americans cleaned themselves by wiping their bottoms with corn cobs or scraping them with mussel shells. Yikes!

- Vikings used leftover sheeps' wool.

- Sailors used the frayed end of an anchor rope.

- Hawaiians scraped their bums with coconut shells.

- Most early Britons wiped their bums with leaves, grass, or balls of hay, unless they were born into royalty, in which case they would have used wool or lace.

- The French used bidets.

- In the late nineteenth century, people in rural America frequently used telephone directories and clothing catalogs.

Are you more likely to catch disease from a toilet seat or a computer keyboard?

Many of us are frightened to sit on public toilets for fear of picking up some horrendous disease, but a study carried out by the University of Arizona revealed that office workers are exposed to more germs from their phones and keyboards than from toilet seats. There are nearly four hundred times as many germs found at a workstation as in a lavatory. There can be as many as 10 million germs on a desk, with more than 25,000 per square inch on the telephone and more than 3,000 per square inch on the keyboard. The average toilet seat contains only about forty-nine germs per square inch. This may be because people rarely touch the toilet seat with their hands, and hands are the main way in which germs are spread. Research has found that in a row of toilets, the toilet in the middle will usually have the heaviest bacterial contamination, whereas the toilet nearest the door will have the lightest.

However, toilet seats can be hazardous in other ways. In 1995, in Ryde, on the Isle of Wight, a man sat on a metal toilet seat and was electrocuted because a faulty cable had caused the toilet seat to become live. In South Carolina, a convicted murderer called Michael Anderson Godwin spent years on Death Row awaiting the electric chair, but eventually his sentence was reduced to life imprisonment. However, when he tried to mend the television in his cell, while sitting on a metal toilet, he bit into a wire and was fatally electrocuted.

Is it possible that the bacteria found in poo could also be on your toothbrush?

When a person has a poo and then activates the flush, the water flushing down the bowl creates a very powerful aerosol of droplets that scatter quite a long distance and may carry fecal (poo) bacteria. Droplets may travel up to 20 feet (6 m) from the toilet, which in many homes is far enough to reach the toothbrush! In a public toilet the flushing mechanisms are often more powerful, so the spray can travel quite far. When the lid is closed the spray will obviously travel a shorter distance, but many of the bacteria will then be found on the underside of the toilet seat.

What is poo?

Poo consists mostly of water, although the proportion of water can vary considerably. For example, the water content in diarrhea is much higher than normal, whereas when we are constipated the poo will be much drier. Water is absorbed from our poo as it passes through the large intestine (which is also known as the colon), so the longer a poo remains inside the body the drier it will become.

Around a third of our poo is made up of dead bacteria that previously lived in the intestines. Fiber, also known as roughage, makes up another third of the poo. Fiber comes from plant foods and is the part of the plant the body cannot digest. The rest of the poo consists of cholesterol, live bacteria, dead cells, and mucus from the lining of the intestine.

Can poo ever be valuable?

In 1972, archaeologists in York, U.K., discovered a very large piece of poo thought to be around 1,200 years old. This Viking poo measured an eye-watering 8 inches (20 cm) by 2 inches (5 cm), and is thought to be one of the largest fossilized human poos ever found. Inside the poo there was evidence of several hundred eggs, which had probably been tapeworm eggs. An average Viking was thought to have hosted up to six hundred tapeworms inside his or her body, because of poor hygiene and a meat-rich diet.

Despite having survived intact for well over a thousand years, the Viking poo broke into three pieces during a school visit to the museum where it was on public display—the Archaeological Resource Centre (ARC) in York. Don't fret, however, as the poo has since been carefully glued back together. The poo is highly collectible and is said to be insured for around £20,000 (about $42,000 U.S).

In the 1960s, an Italian artist called Piero Manzoni exhibited and sold tins containing his own poo. He produced ninety cans of "Artist's Shit," with each tin containing one ounce (30 g) of brown. Manzoni's reason for producing such "pieces of art" was to make a statement about how gullible art buyers were. In 2002, London's Tate Gallery paid £22,300 (about $47,000 U.S.) at Sotheby's for one ounce (30 g) of his canned feces.

Why is poo brown?

Red blood cells live for about three months and are then destroyed in the spleen. The spleen is found in the upper left-hand side of the abdomen, behind the stomach. When the blood cells are destroyed, an orange-yellow substance called bilirubin is released. Bilirubin can also be found in bile, which is released by the gallbladder and is responsible for breaking down fatty foods. Bilirubin combines with iron and waste matter in the intestines and helps to produce that familiar chocolate-brown poo.

Can poo be dangerous?

Most humans have evolved to be repelled by the smell of poo because it transmits disease. "Coprophagy" is the term for the consumption of poo, and this has been observed in a small number of patients suffering with conditions such as dementia, schizophrenia, and depression. Consuming one's own poo is potentially dangerous, as it contains bacteria from the bowels and may also contain the eggs of parasitic worms, neither of which is safe to eat. Eating other people's poo carries the risk of contracting diseases spread through fecal matter, such as hepatitis.

In Tudor times, one of the more unpleasant jobs was that of the "gong scourer," whose task it was to remove the poo from outside toilets and dispose of it. There was often one privy for every twenty households, and it would consist of little more than a lean-to structure with a wooden seat. Richer people would have their own private privies with cesspits. The gong scourer spent a lot of time knee-, waist-, or even neck-deep in poo and would probably have had a couple of boys carrying buckets to help him. He transported a large container on a horse-drawn cart into which he would put the poo. The job was not only disgustingly smelly but also dangerous. There were reports of gong scourers who were asphyxiated by the noxious fumes. Even though scourers were well paid, there were harsh penalties if the waste was not disposed of properly. One scourer made the mistake of pouring poo down a drain instead of taking it outside the city. As a punishment, he was put into his own poo-filled container with poo reaching up to his neck, and a sign was placed nearby describing the crime for which he was being punished.

What is a "floater"?

Floaters are poos that contain a lot of gas or a lot of fat. Sometimes the gases produced by bacteria in our intestines don't have a chance to collect into a large fart bubble but instead get trapped inside our poo. The poo then comes out, perhaps looking a little foamy. If the poo has a lower density than water, it will float.

How do astronauts poo in space?

In 1969, the American astronaut Buzz Aldrin became the first person to have a poo on the moon. He collected the waste in a bag that was attached to his spacesuit. Because of the zero gravity in space, poo would often escape from these bags as astronauts were disposing of it and fly around the shuttle. To deal with this problem, the astronauts' diet contained a minimum of fiber to prevent them from pooing too often.

Astronauts now have special toilets that do not use water and instead work rather like a vacuum cleaner. When they need to poo, the astronauts fasten themselves to the toilet seat and operate a lever that activates a powerful fan. A suction hole slides open, and the poo is sucked away to be collected, compressed, and stored for disposal.

Is it true that King Henry VIII had his bottom wiped by a servant?

Henry VIII liked to eat, and enjoyed a diet that consisted of a lot of meat and fat, and very few vegetables. In later life he had a 54-inch (137-cm) waist and weighed around 740 pounds (336 kg). He was so heavy that he had to be winched into the saddle to get onto his horse. The Tudors believed that the king was so important that everything had to be done for him—including wiping his bum, which was done by the "Groom of the Stool." This position was an extremely prestigious job and well paid. However, it did carry some risks. When Henry was looking for an excuse to get rid of his second wife, Anne Boleyn, in 1536 he accused his Groom of the Stool, Sir Henry Norreys, of committing adultery with her and had him executed.

Why do humans fart?

On average, we fart about fourteen times each day, enough to fill a small balloon. The reason we fart is to expel gas that builds up in our intestines. This gas comes from two main sources: air (comprising mainly oxygen and nitrogen) that we swallow while eating food, and gases produced by the bacteria that live in the large intestine. These bacteria break down certain parts of our food, such as soluble fiber, which have not been digested higher in the gut. When these bacteria consume certain types of carbohydrates (sugars), such as baked beans, they produce a mixture of gases that includes hydrogen, methane, and certain sulfur-containing gases. The smell of farts comes mainly from the sulfur. The more sulfur-rich foods you eat, the more sulfides will be produced by the bacteria in the gut, and consequently, the more your farts will stink. Foods like cauliflower, eggs, and meat are particularly bad for making stinky farts.

Of course, the smell of a fart is only one of the ways in which it can cause offense—there is also the sound. The sound

of a fart is caused by the vibrations of the anal opening, and it depends on the amount of gas produced, the speed with which it leaves the body, and also the tightness of the sphincter muscle of your anus.

In fourteenth-century Europe, it was believed that bad smells could help to ward off the plague. Some people believed that farting into a jar and setting the fart free when the plague was around would help to protect you from it.

People in Japan are particularly sensitive about the embarrassing noises that result from a trip to the loo. To cover up any farts or kerplunks, many Japanese ladies' toilets are designed with specific noise-masking features, such as the sound of a constant flow of running water. Many public toilets in urban areas contain an "Otohime," which is a device that reproduces the sound of a flushing toilet to help drown out any embarrassing noises. The device is activated by pressing a button or by a motion sensor, so the wave of a hand will set it off. Some Japanese women even carry a small torchlike gadget, which also makes the noise of a flushing toilet.

Should we be concerned about skid marks in the pants?

The average person's underwear is said to contain lots of the same types of bacteria as are found in poo, as much as the weight of a quarter of a peanut. Tests have been carried out on the water pumped out of washing machines, and these tests have found that this water contains a significant amount of poo contamination. Clearly then, while our clothes are being washed, it is not only washing powder that is swirling around in the water. This suggests that clothes that have just come out of the washing machine may contain fecal bacteria, so it is recommended that you wash your hands thoroughly after doing the laundry, especially if you are about to prepare or eat food.

Has anyone ever earned money from farting?

In the 1800s, a French baker called Joseph Pujol took to the stage, often wearing clothing that exposed his bottom, and he entertained large audiences with various tunes—all produced by farting. His stage name was *Le Petomane*, which means "The Crazy Farter." His talents included imitating various sounds with his farts and smoking cigarettes with his anus. Using a rubber tube, he was able to blow out a candle 12 inches (30 cm) away, just by bending over and farting.

In 1892, he entertained an audience at Paris's Moulin Rouge. Pujol's act included impressions of the sound of cannon fire, as well as amusing renditions of the farting noises made by, respectively: a nun, a bricklayer, and a bride before and after her wedding night—the joke being that before marriage a woman is very discreet about farting, but after she marries she no longer cares and so lets rip without inhibition.

Can you light a fart?

Yes, but it can be quite dangerous, so make sure you help your gran when she's attempting it. A typical fart is made up of around 59 percent nitrogen, 21 percent hydrogen, 9 percent carbon dioxide, 7 percent methane (although not all farts contain methane) and 4 percent oxygen. Hydrogen and methane are both flammable gases. If the intestinal gases have a higher than normal oxygen content, the display is said to be particularly impressive. If methane is present in your flatus, the flame will burn blue, otherwise fart flames are yellow when ignited.

Ha ha ha
ha ha ha
ha ha ha ha
ha ha...

Why does water not soothe the tongue after you eat hot spicy food?

The spices used in most hot foods are oily. Because oil and water don't mix, when you drink water it just rolls over the oily spices. A far better remedy is to eat some bread or drink some milk, as both of these will help to absorb the oily spices. Milk contains a substance called casein, which binds to the spices, allowing the milk to carry them away. Alternatively, reach for a glass of beer or wine, as alcohol dissolves spices.

Why do we have trillions of bacteria living in our gut?

Bacteria in our gut help to break down some of the carbohydrates found in bread and sugar, which the body uses to produce energy. They can also help the body to absorb calcium, magnesium, and iron, and help fight off harmful microbes that can cause infection. The bacteria in our colon produce the vitamins B and K, which are absorbed and used by the body.

Probiotics are tiny natural bacteria that are added to food

or taken as a supplement to help the digestion of food, and are therefore beneficial to our health. Food with added probiotics will contain billions of "friendly" bacteria. These little friends work alongside our own natural bacteria, called gut flora, to aid digestion. According to research, probiotics can have a beneficial effect for a number of digestive disorders, including helping to control diarrhea and irritable bowel symptoms.

What causes smelly breath?

A healthy tongue will be pinkish in color. However, if bits of food, bacteria, and dead cells are allowed to collect between the little bumps of the tongue, they will build up to form a whitish, smelly coating. This buildup will lead to smelly breath, because the rotting food and bacteria produce unpleasant smells. This coating is not a sign of disease, and is more common in heavy smokers and people who breathe through their mouths rather than their noses. It can be removed using a tongue scraper.

Has it ever been acceptable manners to spit at the dinner table?

Roman diners regularly spat and vomited at the table and had special bowls in which to spew, although they frequently decided to spit or puke on the floor instead. Slaves had the disgusting job of cleaning up the mess. The puke would have consisted of lavish foodstuffs such as dormice drizzled with honey, which was a favorite food of the Romans. The philosopher Seneca (3 B.C.–A.D. 65) wrote, "When we recline at a banquet, one [slave] wipes up the spittle; another, situated beneath, collects the leavings [i.e., the vomit] of the drunks."

In the Middle Ages, it was considered acceptable to spit during dinner, but only onto the floor. It was quite proper to belch at the table, but not in somebody's face. Picking one's nose was fine, too, providing the snot was wiped onto one's clothing or the tablecloth.

In Tudor times, Henry VIII's second wife, Anne Boleyn, would puke between courses. Her maid would hold up a piece of cloth so that she could discreetly vomit into it without offending her guests.

Could a tapeworm be used for weight loss?

Adult tapeworms can grow to up to 30 feet (9 m) long and live in the human body for up to twenty years. The head of the tapeworm has little hooks that fasten onto its host's intestines. While living in your intestines the tapeworm eats your food and so you are hungry all the time, however much you eat. If the infestation is heavy, tapeworms can cause malnutrition and anemia.

In the 1920s, ads appeared promoting "tapeworm pills" as a method for losing weight. Since then, a number of famous women, from opera singers to supermodels, are said to have used them. In the 1950s, overweight opera star Maria Callas (1923–77) lost 62 pounds (28 kg) in one year, after allegedly swallowing a tapeworm. Pills containing tapeworm eggs were on sale in America at the time, but she might also have picked up the parasite from raw steak or liver, both of which she enjoyed eating. Undercooked meat can contain tapeworm eggs or larvae, if the animal itself had tapeworms. Tapeworm eggs are generally ingested through food, water, or soil contaminated with human or animal feces.

A professor of environmental parasitology at Tokyo Medical and Dental University has hosted tapeworms in his gut for more than five years. He believes they have helped him to stay slim. He also claims the tapeworms have cured his hay fever. He suggests that tapeworms can help asthma sufferers and people who are obese. Most people with a tapeworm can be successfully treated with antiparasitic medication.

Can eating carrots help to improve your eyesight?

For generations, people have believed that eating carrots can help to improve the eyesight, and many children have been coerced into eating their carrots for this reason. In fact, carrots can't improve our eyesight, but they can play a role in preventing it from deteriorating if we are unusually deficient in vitamin A. Carrots contain a substance called beta-carotene, which is converted into vitamin A in our bodies. Vitamin A is essential to the retina, found at the back of the eye, and helps to prevent macular degeneration, a common eye disease associated with aging, which leads to vision impairment or even blindness.

In the retina, vitamin A is transformed into the purple pigment rhodopsin, which is essential for vision in dim light. Beta-carotene also acts as a filter, protecting the eye lens, and so reducing the risk of developing cataracts. However, eating excessive amounts of beta-carotene, which is an orange pigment, will result in carotenemia, a condition that turns the skin orange or yellow. Many other foods such as green vegetables are also rich in vitamin A. A well-balanced diet, with or without carrots, provides all the nutrients we need for good vision.

The erroneous belief that carrots will improve night vision seems to have originated during the Second World War. The Royal Air Force spread the rumor that their pilots were improving their night vision by eating carrots, which was giving them a significant advantage over their Nazi counterparts. This was in fact a deliberate lie, designed to cover up the fact that the Allies' real advantage had come about through a new, secret airborne radar system, which was the real reason why the Nazis were being outfought in the air. British Intelligence did not want the Germans to discover this radar system and so promoted the carrot story. One RAF pilot was nicknamed "Cat's Eyes," because of his outstanding night vision and his exceptional score of enemy aircraft shot down at night; his successes were attributed to eating lots of carrots. After hearing this, many people in Britain ensured that they too ate plenty of carrots, believing that this would help them find their way around during the blackouts. At the time, food rationing was severe, but carrots were plentiful.

Which organ can grow back if it is cut in half?

The liver is an incredible organ that is vital for survival. It is the largest gland in the body and is located in the upper right side of the abdomen. The liver carries out more than five hundred functions, including detoxifying the body after we've drunk too much alcohol, regulating blood clotting, and producing bile, which our body uses to break down fats that have been consumed.

The liver is the only organ in the body that can regenerate itself, which means that if part of it is removed, it can grow back to its original size. The liver is made up of little cells called hepatocytes, which produce new cells by dividing into two and are capable of reproducing at a rapid rate. If a patient was to have surgery to have half of his or her liver removed (which could happen as a result of a tumor, for example), the liver would grow back to its normal size within just two weeks.

If you swallow food while standing on your head will it end up in your stomach?

When we swallow food it travels along the esophagus (which is also known as the food pipe) and down into the stomach. The esophagus is a tube around 10 inches (25 cm) long, and its walls contain powerful muscles, which squeeze the food down into the stomach in about two seconds—in a process called peristalsis. Because of these muscles, even if you swallow food while standing on your head, it will still end up in your stomach. However, if you do attempt to eat this way there is a higher risk of choking, so it's not recommended.

Can chewing gum get wrapped around the intestines if swallowed?

There is a widespread, erroneous belief that swallowed chewing gum will stay in the stomach for years or wrap itself around your internal organs, cutting off the blood supply! It is true that chewing gum consists of a fiber that can't be digested by our bodies, but it doesn't get stuck inside us. It takes the same journey as any other food, traveling through the esophagus (food pipe) into the stomach, through the intestines, and out of the anus, probably hiding inside some poo. One benefit of chewing gum is that it increases the production of saliva in the mouth, which helps to protect the teeth from cavities.

Can smoking make your teeth fall out?

The short answer is yes. Research has found that smokers are twice as likely to lose teeth as nonsmokers. Smoking can lead to gum disease, because the chemicals found in smoke affect the gums. For gums to be healthy they need nutrients, which come from the food we eat and are carried by the blood. Smoking reduces the blood flow to the gums, which means that fewer nutrients, such as glucose and vitamin C, reach them. More generally, smoking reduces the body's vitamin C levels by half, and vitamin C is important for helping gums to stay healthy.

Gum disease can lead to the gums' pulling away from the teeth, which can result in teeth becoming loose, or even falling out. A smoker's chances of developing gum disease drop to the same level as a nonsmoker's only after a person has stopped smoking for about eleven years.

Why did women blacken their teeth in Tudor times?

In the late 1500s, many rich people's teeth turned black because they ate too much sugar, which led to tooth decay. Queen Elizabeth I liked sweet, sugary foods, and as a result she lost some of her teeth and others became black from decay. She filled the gaps in her teeth with cloth to make herself look better in public. It became fashionable for women to blacken all their teeth, as this showed that they could afford to buy lots of sugary foods and was therefore a sign that they were wealthy.

Ancient Japanese men and women of the nobility lacquered their teeth black to distinguish themselves from slaves. The dye they used contained substances such as iron filings and tea. They would apply the mixture to their teeth using a brush, until the desired shade was achieved.

What causes an itchy bum?

Itching of the skin in and around the anus is known as "pruritis ani." The intensity of the itching may range from "mildly irritating" to "impossible to ignore." There are various factors that can cause an itchy bum, including thrush, infectious skin conditions, piles, a reaction to certain soaps and creams, or even the dye used in some toilet paper! Not wiping your bum properly can also cause an itchy bum, as the bits of poo that are not properly wiped away can cause the skin to react and become inflamed.

Another potential cause is the threadworm (*Enterobius vermicularis*), which is also known as the pinworm. Threadworms are white and threadlike and can cause severe anal itching. They can be transmitted by contaminated food or drink, and they are more common in children than in adults. Children may get a threadworm infection by getting threadworm eggs on their hands from food, clothing, or toys. From the time a threadworm egg is swallowed it takes about four weeks to become fully grown. The itching begins when the adult female worm lays tiny eggs around the anus and releases a substance that makes the anus very itchy. Thankfully, medicine can be taken that will quickly and effectively kill the worms.

Why do we have an appendix?

The appendix is generally found in the lower right side of the abdomen. It dangles off a wide tube called the large intestine, which is also known as the colon. Many experts believe that the appendix has no significant function. Evidence suggests that our evolutionary ancestors used their appendices to digest tough food like tree bark, but we don't use ours for digestion nowadays. The appendix is rich in special infection-fighting white blood cells, suggesting that it might play some unknown role in helping our bodies to fight disease. Nonetheless, we can live without our appendix without any ill effects, and some scientists believe that, as we continue to evolve, the appendix will eventually disappear from the human body.

Appendicitis is a medical condition that means "inflammation of the appendix," and is usually caused by poo becoming blocked inside the appendix, leading to infection. If no treatment is given, the appendix may burst, leading to a dangerous inflammatory condition called peritonitis. It may also cause an abscess, which consists of fluid and lots of harmful bacteria, to form.

Can kissing help to prevent tooth decay?

Kissing, like chewing sugar-free gum, leads to an increase in the production of saliva, which helps to prevent a build-up of bacteria in the mouth and also helps to get rid of acid. Some bacteria form a sticky coating on the teeth; this is called plaque. When we eat or drink sugary foods these bacteria ingest the sugar and produce acid. Then our teeth will be under attack from this acid for about an hour. The acid is powerful enough to destroy the enamel that covers our teeth, which can lead to sensitivity and decay, causing a hole, known as a cavity, to form. Anything with a pH value lower than five, and therefore more acidic than alkaline, can cause tooth decay. Examples of acidic foods include alcoholic drinks, fruit juices, vinegar, and canned fizzy drinks.

Fluoride in our toothpaste and water helps to prevent tooth decay by making the outer surface, called the enamel, stronger. Fluoride also makes it more difficult for the bacteria in your mouth to produce the acids that lead to tooth decay.

Do your taste buds decrease in number as you get older?

There are little bumps on the tongue called papillae, which help to grip food and contain taste buds. Taste buds allow us to perceive four categories of flavor: bitter, salty, sweet, and sour. The taste buds contain tiny hairs, which are attached to nerves, and these nerves send information to the brain. The brain interprets and collates all this information, and so we become aware of how food tastes. Up to 75 percent of what we perceive as taste actually comes from our sense of smell, as the odor molecules of food give us most of our taste sensation. This is why if we produce lots of mucus, such as when suffering from a cold, we may not be able to taste food.

People are born with about ten thousand taste buds, but as we age, some of the taste buds die. That is why some foods taste stronger to children than adults, and perhaps explains why adults are more likely to enjoy stronger flavors such as mustard or chili. By the time we reach old age, we may have only five thousand taste buds left. Smoking can also reduce the number of taste buds a person has.

Why did people in the late 1700s sell their teeth?

In the late 1700s, the use of sugar became more widespread, leading to an increase in tooth decay. In response to this, it became fashionable for wealthy, hip English people to have tooth transplants. Rotten or damaged teeth would be removed and be replaced with healthy, white teeth, which were often bought from poor, young people, although animal teeth were also occasionally used. A tooth transplant involved taking a healthy tooth from one person and transplanting it into the head of another person whose tooth had just been pulled. Often, if a good fit was not achieved, the next donor would step forward, repeating the process, until there was a satisfactory result. The healthy teeth of executed criminals were also in demand, and there was a consensus that teeth could be legitimately taken from corpses if the person's identity was unknown. There were also stories of teeth being stolen from graves, but these teeth would mostly have been in very poor condition because the corpse would have been rotting. Tooth transplants plummeted in popularity when it was discovered that syphilis could be transmitted in this way.

How long can a person live without food?

Generally, a person can live without food for many weeks, as the body will use its fat and protein stores for energy. Protein stores are found in our muscles, which is why starvation causes muscles to waste away. If a person has a lot of fat stored in his or her body, he or she may live longer than a person who has very little fat.

At the age of seventy-four, Mahatma Gandhi, the famous peaceful campaigner for India's independence, survived twenty-one days of voluntary starvation while allowing himself only sips of water. In 1981, Irish Republican Army inmates carried out a hunger strike to protest the way they were being treated in prison. According to reports, ten of the inmates died having spent between forty-six and seventy-three days without food.

Without water, on the other hand, a person will die within about three or four days, and the size of the person doesn't make much difference to the length of survival.

Is it safe to eat moldy food after the mold has been cut off?

Molds are tiny plants that have threadlike roots that burrow into the foods on which they grow. Molds are mostly visible and grow on many types of food, especially cheese, fruit, and bread. The toxins produced by molds are invisible and can penetrate food. So even if the furry green jacket is removed from food it may still contain toxins, and it may still cause illness if eaten. The safest option is to throw moldy food away.

However, some molds are not harmful as they do not produce toxins, and these are found in certain cheeses with an external mold coating, such as Brie and Camembert, or those with mold running through them, such as Stilton and Danish blue. These molds have been purposely introduced by adding fungi spores to the cheese. However, according to the British

Foods Standards Agency, if mold grows on a cheese that isn't supposed to be moldy you shouldn't eat it.

Is it safe to eat green potatoes?

Green potatoes do not look very appetizing. This is fortunate because the green part of potatoes contains increased levels of toxins called glycoalkaloids, which cause potatoes to have a bitter taste. Eating the green part of a potato can be harmful, and although the most likely outcome is only a mild tummy upset, there have been cases with more serious effects.

In 1979, a large number of children attending a school in London suffered from stomach pain, diarrhea, and vomiting after eating lunch. The illness was caused by high levels of glycoalkaloids in the potatoes that had been served to the children at lunchtime. All of the children recovered, although some required hospital treatment. According to the British Foods Standards Agency severe glycoalkaloid poisoning is very rare, but it's important to store potatoes in a cool, dark, dry place and not to eat the green or sprouting parts. They also advise us not to eat green potato chips.

Is it true you should not swim for an hour after eating?

When you eat, blood is diverted to your digestive system to help process the food. Exercise also creates a demand for blood in the muscles, because during exercise our muscles require more oxygen, and oxygen is carried in the bloodstream. If you eat a big meal and then start exercising vigorously there may be less blood available for the muscles, because it is being used by the digestive system, and this could possibly lead to cramps. This isn't particularly dangerous on land, as you can simply stop exercising and relax the muscles. However, in the water you need to keep moving, otherwise you risk drowning. So, for this reason, it is safer not to swim right after eating.

Is it true the stomach contains a substance that can eat through metal?

A highly corrosive substance called hydrochloric acid is pro-duced naturally in the stomach and helps to digest your food. Hydrochloric acid has many industrial uses, including the pro-duction of batteries, camera bulbs, and fireworks. It's even used to process sugar and make gelatin.

The stomach is a J-shaped sac found in the upper abdomen on the left side of the body. Our gastric juices, which include hydrochloric acid, help to break down food inside the stom-ach. When bacteria, fungi, and other harmful matter enter the stomach after being swallowed, they are destroyed by the hydrochloric acid. The only exceptions are bacteria such as E. coli and Salmonella, which can lead to illness.

The acid in the stomach is so concentrated that if you were to put one drop of it onto a piece of wood, it would burn right through the wood. It would even dissolve an iron nail. Fortunately, the stomach is protected from the hydrochloric acid by a thick layer of mucus.

Do women have more fat cells than men?

Fat, or as it is medically known, adipose tissue, is found underneath the skin. There is also some fat covering each of the kidneys, and in the liver and muscles too. Other locations of fat depend upon whether you are a man or a woman. An adult man tends to carry body fat in his chest, abdomen, and buttocks. An adult woman tends to carry fat in her breasts, hips, waist, and buttocks.

Hormones called estrogen and testosterone are responsible for the position of fat in the body, and it is during puberty that the fat gets distributed. Fat is made up of tiny cells, which can be thought of as tiny individual elastic bags, each carrying a droplet of fat. Fat cells do not tend to increase in number after puberty, but remain roughly constant. However, if we eat too many calories, this can lead to the fat cells' becoming larger, and so we become bigger. Women have more fat cells than men; it has been estimated that men have around 26 billion, whereas women have 35 billion.

If you're starving, is it safe to eat your clothes?

Surprisingly, items of clothing that are made from natural materials such as leather have some nutritional value. One man stranded in the Australian outback survived by eating his leather clothes and drinking water from his car radiator.

In 1994, a fishing boat capsized off the coast of the Philippines, and the fisherman spent many days hanging on to a buoy. When he was finally rescued, people were surprised that he had survived. The fisherman said that eating his underpants had saved his life.

One thirty-five-year-old Indian woman's diet includes cotton clothing. She enjoys eating dishcloths, shirts, and underwear and has done so since she was a small child. Despite her strange eating habits, Bangari is, apparently, a healthy lady.

Why does drinking too much alcohol cause a hangover?

Hangovers can comprise a number of symptoms including headache, diarrhea, nausea, fatigue, and a dry mouth. When we drink alcohol, it enters our bloodstream and causes a gland in the brain to block the production of a hormone called anti-diuretic hormone (ADH). This hormone controls the amount of water in our urine, and as alcohol interferes with the release of this hormone, it means that that the drinker urinates more often. This is why drinkers often have to make frequent trips to the toilet. As a result of all this urination, we lose salts and potassium, which are necessary for proper nerve and muscle functions. When their levels get too low, fatigue and nausea can result.

After heavy drinking, the body sends a desperate plea for water, which is why people with hangovers feel so thirsty. Headaches are thought to be caused by dehydration and also the inflammatory effect of alcohol, which leads to the swelling of blood vessels in the brain.

Alcohol can irritate the stomach's lining and also stimulate the release of hydrochloric acid in the stomach, which may cause our nerves to send a message to the brain, leading to vomiting. The stomach's irritation may also be a factor in some of the other symptoms of a hangover, such as diarrhea and lack of appetite.

Which foods can help cure a hangover?

There is a range of things we can do to help cure a hang-over, which includes drinking water and fruit juice, and eating food such as eggs and bananas. After a heavy night's drinking, our bodies become dehydrated, so drinking plenty of water is obviously beneficial.

Eggs are useful because they contain a substance called cysteine, which is found in most proteins and which helps to break down hangover-causing toxins. Eggs can help to mop up the leftover toxins, and they also give us energy.

When we get drunk we lose potassium through urinating too much. Bananas and kiwi fruits contain potassium, so they can help to replace this loss.

The sugar in fruit is called fructose, and this is also natu-rally found in fruit juice. Fructose helps to restore our energy levels, and studies have shown that it also increases the rate at which the body gets rid of toxins, such as those left over from the breaking down of alcohol. Another benefit of fruit juice is that it contains vitamins, and these may also need topping up after a night on the booze, again because of frequent urination.

What is the "Bristol Stool Form Scale"?

Having lived in Bristol for most of my life, I was proud to discover the "Bristol Stool Form Scale," which is a guide that was developed at the University of Bristol and lists seven kinds of poo formation.

Poo consists mostly of water, although the amount of water can vary significantly. Water is absorbed from our poo as it passes through the large intestine, so the longer a poo remains inside the body, the drier it will become. Some patients may be confused about the precise meaning of constipation and diarrhea, and some may complain of feeling constipated despite having just passed a soft or watery poo. Using this scale can help a patient to understand what type of poo he or she is actually producing, and can help his or her doctor to assess patterns or changes in bowel habit, and, more accurately, diagnose a condition.

The Bristol Stool Form Scale

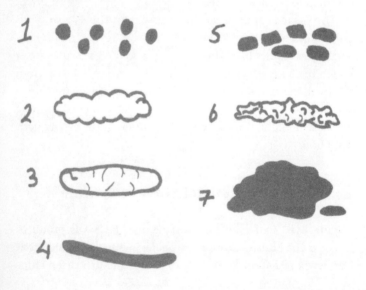

As a rough guide, people who are constipated will pass a Type 1 or 2, and quite infrequently. These stools don't contain much water, which means they will be quite hard. People with diarrhea will pass a Type 6 or 7, extremely frequently. These stools contain a lot of fluid and will have a watery consistency. A "normal" stool should ideally be a Type 3 or 4.

CHAPTER SIX

Cognitive Curiosities

Can a lobotomy help to cure depression?

A lobotomy is a surgical operation that is designed to treat pain or emotional disorders. The beginnings of lobotomy can be traced back to 1848 in America, when an explosion blew an iron bar into the front part of the head of an Irish mine worker called Phineas Gage, causing serious injury. After the accident, the doctors were interested to note that not only did Gage survive, he seemed to have undergone a complete change in his personality as a result. Unfortunately, the change had not been a positive one. Before this accident, Gage was well liked and had been a very soft-spoken and religious man, but afterward he became impatient, fitful, and disrespectful, and used a lot of profane language. It was recognized that following Gage's experience there could be medical applications.

An American neurologist and psychiatrist called Walter Freeman (1895–1972) is known as the father of the lobotomy, and he kept records of the 3,439 lobotomies he carried out during his career. Freeman and his colleague James Watts pioneered lobotomies in America, and often carried them out to pacify aggressive patients suffering with mental disorders. Freeman believed lobotomies worked because they involved severing nerve connections between the frontal lobes of the brain and the thalamus. The thalamus is part of the brain that coordinates nerve messages relating to the senses of sight, hearing, touch, and taste. Freeman thought that the thalamus was the seat of emotion, and something that people who were mentally ill generally had too much of. After the lobotomy was carried out, some of the patients found their conditions improved. However, they often had to relearn how to eat, dress, and use the bathroom, and some died shortly after the operation.

Freeman performed his first lobotomy, in 1936, on a sixty-three-year-old lady called Alice Hammatt, who suffered with depression. Freeman and his assistants drilled six holes into the top of Mrs. Hammatt's skull, and according to Freeman, the patient emerged transformed, able to "go to the theater and really enjoy the play."

Freeman developed another technique to carry out lobotomies, which involved the patient's lying flat with his or her
eyes open. Either a local or general anesthetic was given, and
then an ice pick would be inserted into the patient's eye socket
and tapped with a hammer to break the bone. The tip of the
ice pick would then be pushed about an inch and a half into
the frontal lobe of the brain and moved back and forth. The
procedure was then repeated through the other eye. Afterward, the patient would be left with swollen, bruised eyes.

John F. Kennedy's sister Rosemary was lobotomized by
Freeman at the age of twenty-three. She had suffered from
learning difficulties from an early age, and she was said to have
become increasingly irritable and difficult in her early twenties. In 1941, the twenty-three-year-old Rosemary was living at
a convent school, and, according to the Kennedy family biog-
rapher Laurence Leamer:

> She had begun to suffer from terrible mood swings. She had
> uncontrollable outbursts, her arms flailing and her voice rising
> to a pitch of anger. . . . She sneaked out at night and returned
> in the early morning hours, her clothes bedraggled. The nuns
> feared that she was picking up men and might become preg-
> nant or diseased.

Rosemary's father, Joe Kennedy, consulted Freeman, who recommended a lobotomy to bring Rosemary's behavior under
control. It was rumored that the fact Rosemary had begun
showing a great deal of interest in men, which might have
caused embarrassment to the Kennedy family, was a major
factor in the decision to have the lobotomy carried out.

Unfortunately, the operation was a failure and left her incapacitated. She became incontinent and unable to speak more than a few words, and she would stare blankly at the walls for hours on end. Until her death, at the age of eighty-six, Rosemary needed full-time care.

In 1967, Freeman performed a lobotomy on a female patient. During the operation, he accidentally severed a blood vessel, and the patient died of a hemorrhage three days later. This effectively brought his career to an end, and he died from cancer in 1972, aged seventy-six.

Why does eating ice cream cause "brain freeze"?

"Brain freeze" is the feeling you get when you eat or drink something cold, such as an ice cream, too quickly in hot weather, and you get a sharp, stabbing pain in the forehead. The reason for this is that there are lots of very sensitive nerves found above the roof of the mouth that help to protect the brain. When these nerves feel something cold they may overreact, perhaps as a warning in case the coldness could affect the brain. As a result, they send messages to the brain, telling it, "warm up." In response, the blood vessels in our head widen, bringing more blood, and therefore more heat, to the head. This rapid swelling of the blood vessels, which contain delicate nerve fibers, is what causes the pain associated with an ice cream headache.

One way to get rid of the pain quickly is to press your

thumb or tongue against the roof of your mouth, behind the front teeth, as the warmth will help to stop the pain.

Is it true that Coca-Cola used to contain cocaine?

Cocaine is a drug extracted from the leaves of the coca plant. It is a powerful brain stimulant and one of the most addictive drugs known to man. Cocaine causes immediate euphoric effects, which include increased energy, reduced fatigue, and increased mental clarity.

In 1886, an American pharmacist called John Pemberton invented Coca-Cola. It was sold only in drugstores and contained a large amount of cocaine. Cocaine is of course an illegal drug nowadays, but this wasn't the case back then. During these initial years, Coca-Cola was marketed as not only a refreshing drink and pick-me-up, but also a "brain tonic." It was advertized as a cure for headache, depression, hysteria, and muscular aches and pains. Pemberton rejigged the recipe for Coca-Cola a number of times, and the amount of cocaine included in the drink gradually declined to a mere trace, before being removed completely by 1903.

Coca-Cola wasn't the only consumer product to contain cocaine. In the late 1800s and early 1900s, Americans could buy dozens of products containing cocaine, heroin, or opium without a prescription at their local pharmacy or by mail order. Products that contained cocaine included "Cocaine Toothache Drops" and "Anglo-American Catarrh Powder." "Cough Cure" contained a mixture of marijuana and morphine, and

"Dr. James Soothing Syrup" contained heroin. Even babies weren't safe. "Dr. Moffett's Teethina Teething Powders," which were used to help teething babies, were packed full of opium.

Why do we yawn?

Humans are not the only creatures who yawn—there are a number of animals, fish, and even insects (ants, specifically) that also appear to yawn. Many of us yawn when we are in a situation that is boring or if we are tired. Some experts think that yawning helps to stimulate and awaken the body in situations where we have to stay awake or alert. This theory helps to explain why people who drive late at night on motorways yawn a lot.

In one experiment, Italian researchers videotaped premature infants in intensive care units. They found that the babies yawned before sleep, and also as they awoke. These researchers concluded that yawning indicates a changing state of arousal in the body, such as the shift from sleep to waking.

Studies show that, contrary to expectations, yawning isn't the body's way of increasing its oxygen supplies. An experiment carried out on a group of psychology students at an American university showed that the subjects yawned just as much in a room containing air with lots of oxygen as they did in a room with much less oxygen.

There is some link between yawning and certain diseases, although it is unclear why. Excessive yawning is associated with a variety of conditions, mostly concerned with the brain, such as epilepsy and multiple sclerosis. People with schizophrenia, on the other hand, yawn very little.

Why is yawning contagious?

Yawning is definitely contagious. In fact, even reading, hearing, or thinking about a yawn can cause us to start yawning ourselves. Some experts think that yawning may have developed as a means of communication. It may, for example, be a way of signaling to others that it's important to remain alert and stay awake in a certain situation.

Another theory is that we have evolved to yawn at the sight of others yawning because our early ancestors used it as part of their social behavior and as a way to help build a bond with the rest of the group. When one member of the group yawned, it may have been a sign that it was time to sleep, so the rest of the group might have yawned in response to show that they agreed. Babies are unaware of these signals and so they don't yawn contagiously until they're about one year old.

Why do legs "fall asleep"?

It's a strange sensation waking up in the morning with a limb that has "fallen asleep." Then, the "pins and needles" arrive, and these are a sign that the limb will soon be working properly again. But why does this happen?

The nerves in our bodies are responsible for every movement we make and every sensation we feel. Nerve messages are continually being sent to and from the brain. Every movement we make requires a message to be transmitted from the brain, through the spinal cord, to the area of the body we would like to move. Arms and legs often fall asleep when they are squeezed under the body, as when we kneel on one knee, because this causes the nerves in the affected leg to be temporarily squashed. For body parts to move, our nerve messages must be able to pass freely through the nerves. If our nerves become so squashed that the messages cannot be passed through them properly, the limb becomes unable to receive or send messages, and as a result it may become numb and fall asleep. The limb may remain numb for a while, but as the nerves begin to stretch back to their normal shape they will begin to transmit messages once again. This leads to the sensation of pins and needles, until the nerves in the limb function normally again.

Do intelligent people have bigger brains?

Scientists originally thought that the whole brain was in charge of all bodily activities and so assumed that a person with a large brain must therefore be intelligent. But it was later discovered that people with manual jobs had brains as large as professors, and that criminals had brains no smaller than those of law-abiding people.

Albert Einstein (1879–1955) is universally acknowledged as a genius, but his brain was only of average size. However, Einstein's brain was different from normal brains in one key respect. When Einstein died, in 1955, Princeton Hospital pathologist Dr. Thomas Harvey performed the autopsy and removed his brain. Afterward, the brain went missing, and its whereabouts was a mystery until it was discovered that Dr. Harvey had kept the sliced-up brain at his home in Kansas. The brain had been stored in his office, in two jars, inside a cardboard box labeled "Cider." Then, in the 1980s, for reasons never made clear, Thomas Harvey sent out parts of the brain to scientists and researchers around the world.

Studies on the brain revealed that Einstein had had an unusually large inferior parietal region, which is the part of the brain responsible for mathematical thought—Einstein's inferior parietal region was 15 percent wider than that of other brains. Einstein's brain was also unusual in that it didn't contain a groove called the sulcus, and this may have been a key factor in Einstein's unusually high intelligence, as the absence of this groove may have allowed nerve cells on either side to pass messages more rapidly or easily. However, many scientists who have reviewed these studies are unconvinced, stating that the physical differences in Einstein's brain compared to average brains wouldn't necessarily account for his incredible mathematical abilities.

Why do we "jump" when we fall asleep?

The jerking movements we experience while falling asleep are known as hypnic jerks and can be associated with a feeling of falling. There are a number of theories as to why they happen. One theory suggests that they may be a protective reflex. As we go to sleep, our muscles relax and eventually become quite loose. This loosening of the muscles may be interpreted by the brain as a sign that we are falling (even though we are lying down in bed), so the brain sends a message to the muscles to tell them to tighten up to help us stay upright. Hypnic jerks are more likely to happen when we are overtired or when we have been using stimulants such as caffeine.

Why do paper cuts hurt so much?

Our skin contains millions of sensitive nerve endings, of which there are particularly high concentrations in our lips and fingertips. These nerve endings can sense heat, cold, pain, and pressure, and they transmit these messages to the brain. Because our fingertips have a particularly high number of nerve endings, a paper cut to the fingers will generate a greater response than a cut to another part of the body, which is why it will cause more pain.

Can the brain feel pain?

An adult brain contains around 100 billion nerve cells. The largest section of the brain is called the cerebrum, and this consists of two halves known as the cerebral hemispheres. Confusingly, the left hemisphere controls the right side of the body, and the right hemisphere controls the left side of the body.

Although the brain registers pain felt in the body, it can't actually feel pain itself as it doesn't have necessary pain receptors on its surface. Certain arteries and veins in the brain also lack the ability to sense pain. For this reason, a patient undergoing brain surgery requires anesthesia for the skull, but not for the brain itself. Surgeons can operate on the brain while the patient is fully awake and able to talk and answer questions. Some people undergoing brain surgery have even volunteered to undergo experiments in which their exposed brain is stimulated with electrodes during surgery. When not under general anesthesia, they can even report the resulting sensations to the researchers.

The pain of a headache is mainly caused by pain-sensitive structures outside the brain, such as the many nerves that serve the blood vessels and muscles of the scalp, face, and neck.

How did ancient Greeks use eels to relieve pain?

The electric eel is capable of generating powerful electric shocks. According to Galen (A.D. 129–c. 216), who was a famous Greek doctor during the time of the Roman Empire, ancient Greeks sometimes applied an electric eel to the heads of patients to alleviate headaches and to the body to numb pain. The Greeks and Romans also used eels to help ease the pain of gout by making the patient stand on one or a number of electric eels until his or her foot became numb.

Why did ancient Egyptians scoop out the brain when preparing a mummy?

Ancient Egyptians believed in an afterlife, and they thought that one way to achieve immortality was to preserve a person's corpse by wrapping it up in linen bandages in a process called mummification. They went to great lengths to preserve bodies, usually those of wealthy Egyptians, and placed items such as jewels, tools, food, and even pets with the bodies, as it was thought these items would be needed in the afterlife. Embalmers spent a great deal of time ensuring the bodies were thoroughly bandaged to protect them from the decaying effects of the environment. However, the bandages could not stop the effects of the bacteria contained within the bodies, and when the embalmers realized that the corpses were being damaged from the inside they began removing the internal organs to help ensure proper mummification. After the removal of the organs, they would fill the resulting space with sawdust and rags, among other things.

To remove the brain, embalmers chiseled through the bone of the nose and inserted a long iron hook into the skull to scrape out the brain. A long spoon would then be used to scoop out any last remnants, and afterward the skull would be rinsed out with water. The Egyptians tried to preserve most organs by coating them in resin and wrapping them in linen strips, which would then be stored in decorative pots in the tomb. They didn't try to preserve the brain, as they didn't think it was an important organ and assumed you wouldn't need it in the next life.

Could near-drowning cure mental illness?

In the 1600s, a Belgian physician and chemist called Dr. van Helmont immersed his mentally disturbed patients in water, to the point of nearly drowning, as he believed it would extinguish a "too violent and exorbitant form of fiery life." He had discovered this cure after he noticed that:

> many Fools . . . who accidentally fall into water and are dragged out for dead and remain so for a long time . . . that when a dagger's sheath with the tip cut off is thrust in their fundamen, [i.e., anus] and someone blows through it till water gushes out of the drowned person's mouth, they are not only restored to life . . . but also to the full use of their understanding.

In the 1700s, an American physician called Dr. Willard ran an asylum and decided to try the same therapy. He had a tank into which the patient, enclosed in a coffinlike box with holes, was lowered. The patient was kept there until bubbles of air could no longer be seen and was then taken out and revived. The idea was that if the patient was nearly drowned and then bought back to life, the disease would somehow disappear. Of course, these methods did not work; if the patient appeared calmer after these "treatments," it was probably due only to shock.

In the 1800s, some people believed that madness could be shaken out of people. Dr. Benjamin Rush (1745–1813) recommended putting patients into a cage on a pulley and spinning them around and around for hours. Not surprisingly, when the cage was spun around at top speed it caused nausea, vomiting and many other unpleasant sensations. When the treatment was completed the patient would appear less manic, but also very pale and ill-looking. Dr. Rush believed that spinning would help mentally ill patients because it reduced the force of the blood flow to the brain, lessened the pulse rate, and relaxed the muscles of the body. He also thought that vomiting would lead to a more healthy blood circulation.

Could pulling out teeth help to cure mental illness?

In modern times, although the exact cause of most mental illnesses is not known, research shows that many of these conditions are caused by a combination of biological, psychological, and environmental factors. Some mental illnesses have been linked to an abnormal balance of chemicals in the brain called neurotransmitters, such as serotonin. Neurotransmitters help nerve cells in the brain to pass messages to one another. If these chemicals are out of balance or are not working properly, messages may not make it through the brain correctly, leading to symptoms of mental illness. Doctors may prescribe antidepressant medication, which helps to increase serotonin levels in the brain.

In the early 1900s, an American doctor called Henry Cotton (1886–1933) carried out surgery to remove infected body parts, believing it would help his patients' mental illness. "The insane are physically ill," he proclaimed. Dr. Cotton thought that bacteria could spread from an infected site and travel around the body to the brain, causing mental illness and probable death.

Dr. Cotton would often begin by pulling out the teeth, one by one. If that didn't cure the patient's mental illness, he would then proceed to remove organs. He removed organs such as the tonsils, gallbladder, stomach, large intestine, spleen, ovaries, and testicles. Unsurprisingly, more than a third of his patients died.

When Dr. Cotton himself became mentally ill, he had his own teeth surgically removed and then returned to work. Before he retired, in 1930, he'd ordered 11,000 teeth to be pulled from patients at his hospital. Amazingly, he was widely considered to be a pioneer by the medical authorities.

Can anything live inside the human brain?

As shocking as it may sound, there is one unpleasant organism that can reside in the brain and cause the rapid death of its victim.

The organism is called *Naegleria* (nuh-gleer-e-uh) *flowleri amoeba*, and it is found all over the world. It often occurs in warm, stagnant bodies of fresh water such as lakes, rivers, hot springs, and unchlorinated swimming pools. It cannot be contracted by simply drinking water or wading in it; the water must go up a person's nose. The amoeba enters the body through the nose, usually when a person is swimming underwater or diving. From the nose, it climbs along nerve fibers, through the skull, and into the brain. These amoebae love the warmth of the brain and will multiply in their millions until the victim drops dead, usually within just three to seven days of infestation.

There have been around two hundred recorded cases of *Naegleria* infection worldwide in the past forty years. Children are thought to be more at risk, because they have weaker immune systems. However, although the infection is rare, if you do decide to swim in warm, fresh water, remember to wear a nose clip!

Can alcohol improve your snooker?

The famous Canadian snooker player Bill Werbeniuk (1947–2003) drank up to 25 pints (14 liters) of lager each day, and was said to drink 8 pints (4½ liters) before a match, to help "steady his nerves."

He suffered from a condition called "familial benign essential tremor." This nervous disorder is characterized by shaking of the hands and arms, and sometimes other parts of the body too. Its cause is unknown.

In the 1980s, Werbeniuk lived in England and drove

around in a converted bus, which contained all the facilities he needed, including lager on tap. His excessive alcohol consumption obviously contributed to his large frame—he weighed more than 280 pounds (128 kg). During one televised world championship at the Crucible Theatre, he attempted to stretch across the table, but found it difficult because of his size. As he stretched, he let out a loud fart. Everyone tried not to laugh. He then turned to the audience and said, "Who was that?" and the Crucible erupted with laughter.

At his last professional match, in 1990, he said, "I've had twenty-four pints of extra-strong lager and eight double vodkas and I'm still not drunk!" At fifty-six, he died of heart failure.

Who was the maddest monarch?

Down the centuries, various members of the European monarchies have become renowned for their strange behavior, and quite a few suffered with mental illness, as the following will demonstrate.

When his brother Ludwig II died in 1886, Otto became the king of Bavaria—although he probably didn't realize it. Otto claimed that spirits who lived in his dresser drawers advized him that in order to remain healthy, he should shoot a peasant every day. He decided to take pot shots at the peasants who worked in his gardens, but when his staff numbers began to dwindle, one servant was given the task of loading the king's gun with blanks. After this, when the peasants were "shot," they would pretend to fall down and die.

Queen Juana of Spain (1479–1555) adored her handsome, womanizing husband Philip. In 1506, at the age of 28, Philip died; it was rumored that he had been poisoned. She could not bear to be parted from the corpse, and continually caressed the body until his coffin was buried. Five weeks after Philip's death, rumors began to circulate that the body had been stolen. To resolve this question, officials dug up the body, and unwrapped the corpse. The rotting body was still in the coffin, and Juana began kissing its feet. From that point on, she took the coffin with her wherever she went, frequently opening it to gaze at the rotting remains of her husband.

Charles VI of France (1368–1422) was convinced he was made of glass and insisted on iron rods being inserted into his clothing, to stop him from breaking. He didn't like traveling by carriage, in case the vibrations caused him to shatter into pieces. He also prowled the corridors of the palace, howling like a wolf, and refused to bathe for months on end. This displeased his wife, Queen Isabeau, who requested that a young woman called Odette de Champdivers take her place in bed. Every night for 30 years, Odette wore the Queen's clothes in the royal bed and Charles never even noticed the deception.

Is laughter really the best medicine?

When we laugh, our rate of breathing quickens, increasing the amount of oxygen in the blood, which helps healing and improves the circulation. This increased oxygen also causes the blood vessels close to the surface of the skin to expand, which is why people go red in the face when they laugh. It can also lower the heart rate and burn up calories. It is reported that a hundred laughs will give the body a workout equivalent to a ten-minute session on a rowing machine.

Laughter also stimulates the production of chemicals in the brain called endorphins, which are the body's natural pain-killers; endorphins also have the pleasant effect of making us feel happy. They are similar in composition to the drugs morphine and heroin, and have a tranquilizing effect on the body. Endorphins also have a beneficial effect on our immune system and so help defend the body against illness, which may explain why miserable people become ill more often than happy people.

In 1964, Norman Cousins was diagnosed with ankylosing spondylitis, a form of spinal arthritis. He was told that nothing could be done to help him and that he would die in agony. He believed that laughter would help his condition, so he rented every funny film he could find, including *Airplane*, the films of the Three Stooges, and many others. He watched these films over and over again, and laughed as hard and loud as possible. After six months, doctors were astounded to find that his illness had been cured, and he no longer had the disease. In the 1980s, as a result of Norman Cousins's experiences, some American hospitals introduced a "Laughter Room," which was

Ha ha ha ha
ha ha ha
ha ha ha
ha ha...

filled with joke books and comedy films, and hosted regular visits from comedians. Patients were given thirty- to sixty-minute sessions each day, which resulted in dramatic improvements to their health. These patients required fewer painkillers and became easier to deal with.

Is body language the same all over the world?

Because of the global influence of American films and television, Western body language has become increasingly widespread nowadays. But, traditionally, different gestures have had very different meanings throughout the world. In Europe and North America, making an "O" gesture with the fingers and thumb generally means "OK," but in countries like Russia, Turkey, and Brazil, it is a sexual insult, which insinuates that the person being referred to is gay.

In many countries, giving the "thumbs-up" gesture means

that something is OK or good, and is also used by hitchhikers. However, in Greece it can mean "Up yours!"

In most countries, nodding your head means "Yes," but in Bulgaria it means "No." In Japan it means "Yes," but not necessarily "Yes, I agree with what you're saying"; it can also mean "Yes, I understand what you're saying but I don't agree with you." In Arab countries a single, upward head movement signifies "No." In India, the head is rocked from side to side to mean "Yes."

In Saudi Arabia, it is fairly common for men to hold each other's hands in public, purely as a sign of mutual respect. In the U.K., people would probably assume they were gay.

Westerners and Europeans regard it as good manners to blow their nose into a tissue or handkerchief, whereas many Asians would find this habit disgusting. Asians would be more likely to spit or snort the offending substance, as they believe this is more hygienic than blowing the nose into a tissue.

How can you tell if someone is lying?

There are a number of nonverbal clues that can help you to work out whether someone is lying. When lying, both men and women increase the number of times in which they swallow saliva (gulp), but this is more noticeable in men, because of the Adam's apple. Also, when people are lying, they tend to touch their faces more often.

In 1998, U.S. president Bill Clinton denied having had "sexual relations" with former White House intern Monica Lewinsky. As he made his televised statement, "I did not have sexual relations with that woman!" he not only gulped, he also rubbed his nose. According to research, when people lie, chemicals are released that cause tissue inside the nose to swell. The nose expands with blood—this is known as the "Pinocchio Effect." The nose may also become itchy, so liars will often rub their noses to get rid of the itch. During that televised statement, Clinton only ever touched his nose when he allegedly was telling a lie.

Many liars will maintain eye contact while lying to appear convincing, but they may rub their eyes to avoid looking at the person to whom they are lying (alternatively, of course, they could just have an itchy eye!). Other possible signals include jaw clenching and tightening of the lips. Stuttering and using lots of "um" and "ah" words can also indicate that a lie is being told. Liars are also more likely to cross their arms and legs.

Some people may cover their mouth with their hand when lying, as if their brain is subconsciously trying to stop them telling more lies. They may even try to mask the meaning of this gesture, by giving a fake cough. Similarly, if the listener covers his or her mouth, this may indicate that the listener thinks the speaker is hiding something. When people lie, they sometimes increase their lower body movement through actions such as shuffling their feet.

How can body language indicate that a person fancies you?

If a person is attracted to you, he or she will probably hold eye contact slightly longer than usual, and if you look into his or her eyes his or her pupils may be dilated. This widening of the pupils is due to the release of a hormone in the body called adrenalin. A psychologist called Neisser found that when shown two pictures of the same woman, with the pupils artificially dilated in one picture, people rated the version with dilated pupils as more attractive, even though they were unable to explain why.

A person who is romantically interested in you may subconsciously point at you with a foot or knee, which is effectively a signal that says, "I would like to go in this direction." People who fancy someone often touch themselves where they would actually like to touch the other person, or would like to be touched themselves—for example, stroking their legs.

If a woman is attracted to someone she may toss her hair, give sideways glances, or smile. If she is stroking an object such as a wineglass stem or some other phallic-shaped object,

it could indicate that she is thinking about sex. Wiping imaginary dust off the other person's clothes, leaning forward toward the person and standing or sitting closer to the person are all signs of attraction. On the other hand, a woman who is not interested in a man will fold her arms across her chest and cross her legs away from him.

If a man is sexually attracted to someone he may hold his legs apart, subconsciously displaying his genitalia, or stand with both thumbs in his front pockets or belt, using his fingers to draw attention to his crotch by framing it.

Is it possible to forget your accent?

There is a rare condition called Foreign Accent Syndrome, which results in people suddenly developing a foreign accent. It usually occurs after a person has suffered a stroke or some other type of head injury. The majority of sufferers adopt German, Swedish, or Norwegian accents, but the reasons for this are unknown.

In 1999, a Yorkshire woman called Wendy Hasnip, who was forty-seven years old, suffered a stroke, which resulted in her adopting a French accent. She had only ever been to France once and could not even speak the language.

In the same year, American Tiffany Roberts, who was fifty-seven, also suffered a stroke, which left the right side of her body paralyzed. At first she couldn't speak at all, but after months of therapy she was once again able to speak. However, instead of her usual Indiana twang, she began to speak with an English accent. Her new accent was a mixture of cockney and West Country, and she even used peculiarly English words such as "loo." She also spoke with a much higher pitch than before. Her family and friends had problems understanding her, and people accused her of lying when she said she was born and bred in Indiana.

Perhaps the strangest case was that of a Norwegian woman who fell into a coma after being hit on the head by shrapnel during an air raid in 1941. When she woke up, she spoke with a strong German accent. On hearing this accent, her neighbors were outraged and ostracized her.

What is a placebo?

The word "placebo" refers to any medication that the patient believes to be therapeutic, but which, in reality, has no physiological effect; a placebo is often simply a pill made up of sugar or starch. Nonetheless, patients who take placebos often demonstrate a genuine improvement in their health. This shows that mental attitude can play an important role in the treatment of conditions such as pain, depression, and heart ailments. Trials have repeatedly shown that simply believing that a medicine will make you feel better can often have the desired effect—even if the medicine itself is fake. One theory for this is that the "placebo effect" serves to stimulate the release of endorphins, the chemicals in the body that relieve pain.

In America, a test was carried out to see if patients suffering with arthritic pain in their knees could be helped by receiving "placebo surgery." In placebo surgery, no actual surgical procedure takes place, but the patient is led to believe that it has. Ten patients were scheduled to have a standard operation, but half of them underwent placebo surgery instead. In these five cases, the surgeon simply made cuts into the patients' knees three times with a scalpel to create the expected incisions and scars. After six months, all of the patients, including the placebo group, reported significant reductions in pain in the affected areas.

Asthma is a condition in which there is tightening of the air passages to the lungs, which can cause wheezing and breathing difficulties. An asthmatic may use medicine, such as a bronchodilator, which opens up the main air passages to the lungs and helps breathing. In a study of asthmatics, researchers found that they could produce widening of the airways just by telling people they were inhaling a bronchodilator, even when they weren't. Doctors have even managed to cure warts simply by painting them with a brightly colored dye and promising the patients that the warts would be gone when the color had worn off.

Could human beings ever become immortal?

As crazy as it may sound, some scientists do believe that humans could soon become immortal. In the meantime, our only hope of achieving immortality is by being frozen after death and waiting until the technology becomes available to restore us to life.

Cryonics is the term for the practice of freezing corpses, often by using liquid nitrogen. People pay a lot of money for this process in the hope that one day they will be resuscitated, cured of the condition that killed them, and once again given life. Some people choose to have their whole body frozen, while others choose the cheaper option of freezing the head only, in the hope that one day the technology will exist to attach the head to a new, healthy body.

In 1962, physics teacher Robert Ettinger published a book called *The Prospect of Immortality*, in which he proposed that people could have their corpses frozen in the hope that they might one day be revived by more advanced technology. Inspired by Ettinger's writings, in 1972 Fred and Linda Chamberlain decided to set up the Alcor Society for Solid State Hypothermia. The name was changed to the Alcor Life Extension Foundation in 1977. The company is based in Arizona. In 1976, Alcor performed its first human cryopreservation. Currently, Alcor patients are stored under liquid nitrogen at a temperature of minus 321°F (minus 196°C). Liquid nitrogen is used because it is inexpensive and reliable. At the end of August 2006, Alcor had 809 members and 74 patients in cryopreservation.

One of Alcor's patients is American grandmother Anita Riskin. When she was terminally ill with cancer, she decided she wanted to be cryonically frozen after death. She believed that she would be brought back to life in the future, and that there would be a cure found for the cancer from which she was suffering. The process cost her about $165,000. When Anita died at age sixty there was no autopsy, no embalming, and no hearse. Instead, body freezers packed her still-warm body in ice, before taking her straight to Alcor.

Within hours, the blood was flushed out of her body, a hole was drilled through her skull, and probes were inserted into her brain. Her body temperature was taken down to minus 321°F (minus 196°C). Her body was then zipped into a sleeping bag and lowered headfirst into a 9-foot (3-m) tank of liquid nitrogen next to two other corpses, a few pets, and a large number of disembodied heads.

What is "Dr. Strangelove syndrome"?

This condition is named after the main character of the 1960s film *Dr. Strangelove*, who was played by Peter Sellers. In the film, the wheelchair-bound Dr. Strangelove is a mad German scientist, whose eccentricities include a severe case of "Alien Hand syndrome"—he has an unruly hand, clad in a black leather glove, over which he appears to have little control. This out-of-control hand alternates between doing Nazi salutes and strangling his own neck.

In real life, Alien Hand syndrome (which is now also known as Dr. Strangelove syndrome) is a rare condition caused by damage to certain parts of the brain, which results in a person's hand acting independently and taking on a life of its own. For example, the misbehaving hand may do the opposite of what the other hand is doing. So, if a person is trying to tie a shoelace with one hand, the other will try to undo the lace. If one hand pulls up the person's trousers, the other will pull them down.

Sometimes sufferers are not even aware of what the rogue hand is doing until it is brought to their attention, and then they often become angry with it, fighting or punishing it in an attempt to control it. The two hands will sometimes get into a fight. The rogue hand may become aggressive and slap, pinch, or punch the sufferer, and there have even been cases of the rogue hand's trying to strangle its owner. The sufferer will often sit on the hand to try to control it, but eventually it gets loose and starts again.

Why can't we tickle ourselves?

Tickling relies on the element of surprise to generate the feelings of panic, unease, and laughter that follow. When we tickle ourselves, the tickle doesn't have that surprise factor because a part of our brain called the cerebellum is able to sense when the tickle is coming and informs the rest of the brain to ignore the sensation.

The cerebellum deals with movement and also predicts what kinds of sensations to expect. In general, the brain pays little attention to expected sensations, such as the feeling of your fingers tapping on a keyboard. However, unexpected sensations cause a powerful reaction—just think how you'd jump if someone crept up behind you right now and tapped you on the shoulder!

This difference between our bodies' reactions to expected and unexpected sensations is probably a survival mechanism that evolved to ensure our brains paid proper attention to important matters such as detecting predators.

Ha ha ha ha ha ha ha ha ha ha ha ha...

Could a head remain alive after decapitation?

After the French Revolution, during the time of the guillotine, doctors carried out studies to find out if a person's head could remain alive after being decapitated. To research this, the doctors would ask people who were about to be executed to try to keep blinking their eyes after their heads had been chopped off. According to reports, some of the heads continued to blink for up to thirty seconds after being cut off, although it is unclear whether this was because of the person's blinking on purpose or was simply an involuntary reaction.

In 1587, Mary Queen of Scots was executed for her involvement in a plot to kill Queen Elizabeth I. The crowd was shocked when they witnessed her execution. According to reports, Mary's executioner must have been either inexperienced or simply having a bad day, because he tried to chop her head off three times without success before finally using his sheath knife to finish the job. After the first chop, Mary gave a deep, long groan and it was obvious that she was suffering a great deal of pain. When Mary's head was finally severed, her lips are said to have stirred up and down for almost fifteen minutes. When the executioner tried to pick up the head, it rolled away, and he was left holding an auburn wig. Mary's years in prison had seriously compromised her health and beauty, and by the time of her death her natural hair was short and gray.

Then, to everyone's disbelief, the headless body began to move! Mary had had a small dog as a companion, which she had hidden under her gown and taken to the execution. The dog had become stuck in her clothing and was trying to break free. Its wriggling had caused the jerking movements witnessed by the crowd. The dog was apparently in shock and covered in blood following the beheading. It died a few days later.

A French woman called Charlotte Corday (1768–93) was sentenced to the guillotine after stabbing and killing revolutionary leader Jean-Paul Marat, an activist in the French Revolution, in his bath. Marat spent a great deal of time in the bath, as he suffered with a painful skin condition. After Charlotte's head was chopped off, the executioner's assistant picked it up by the hair and slapped Corday's cheek. Many witnesses stated that Charlotte's face flushed red with anger and that she also sneered at her assailant!

CHAPTER SEVEN

Ocular Obscurities

Can anything live inside our eyes?

It's horrible to think that anything could live inside a person's eyes, but there are certain parasitic worms that can take up ocular residence. An eye condition called "river blindness" can be caused by two types of insect: the blackfly and the buffalo gnat, which live on rivers in Central Africa and South America. These insects bite their victim's skin and leave behind the larvae of a parasitic worm, which is then carried in the bloodstream. When the larvae become adults they breed, releasing millions of offspring that spread throughout the body, carried by the blood. Some victims of this infection may not notice any symptoms, while others may suffer from skin rashes, eye lesions, or bumps under the skin. When the larvae develop they can set up home in every part of the eye apart from the lens, causing inflammation, bleeding, and in some cases, even blindness.

Another potential invader is the botfly. This insect lays its eggs on the bodies of other living creatures, such as mosquitoes, which can then pass these eggs on to humans. In one example, a five-year-old Central American boy felt irritation in his eye, followed by a swelling. The swelling got redder and bigger, and eventually the boy was taken to see a doctor, who initially suspected a tumor or cyst. The boy underwent surgery in Honduras, during which the problem was found to be the late-stage larva of the human botfly. The larva was nearly ¾ inch (2 cm) in length and was removed under general anesthesia through a small incision in the conjunctiva: the thin, clear membrane that covers the white of the eye. You can even find pictures of this operation on the Internet—believe me, it's gross!

How do "beer goggles" work?

The "beer goggles" effect refers to the way in which we often find people more attractive after we've had a few alcoholic drinks. Many of us have worn beer goggles at some point in our lives. Perhaps at a party your eyes met across a dark, smoky room, and even though he or she was wearing a wig and had one eye and sweet corn–colored teeth—there was something about him or her that you couldn't resist. But why, the next day, did you regret the kissing and fondling? Presumably, you had been wearing your beer goggles.

Research has shown that it is not just alcohol that is responsible for this effect; additional factors such as the amount of light in a room, the drinker's own eyesight, and the amount of smoke in the air all play a part. A survey showed that 68 percent of people had regretted giving their phone number to

someone while under the influence of alcohol and later realized they were not attracted to him or her.

A study carried out in Scotland involving eighty students
confirmed that the beer goggles effect does occur after drinking alcohol. The students were shown color photographs of
men and women and were asked to rate their attractiveness.
Half of the group had drunk either two pints of lager or two
and a half glasses of wine. Overall, the tipsy students gave the
people in the photographs higher ratings than did the sober
students. According to the study, men and women who had
drunk a moderate amount of alcohol found the faces of
members of the opposite sex 25 percent more attractive. Interestingly, this effect was not confined to the good-looking photographs—even photographs that were rated as being
relatively unattractive were given higher ratings by the group
who had drunk alcohol.

This effect is believed to be caused by alcohol's stimulating
a part of the brain called the nucleus accumbens, which deter-
mines the attractiveness of a person. This area will be naturally
stimulated when you look at someone very appealing. Drinking alcohol stimulates this area, fooling you into thinking that
someone is more attractive than he or she actually is, and this
increase in perceived attractiveness seems to be directly pro-
portional to the amount of alcohol consumed.

Why do onions make us cry?

Onions contain sulfur compounds, which are an irritant to
both the eyes and the nose. Cutting into an onion releases

these sulfur compounds into the air, and when they come into contact with the water in our eyes they produce sulfuric acid, which causes irritation. Our eyes then produce tears to help flush away the irritant. Rubbing our eyes only irritates them further because, having touched the onion, our fingers will now have some of the sulfur compounds on them.

British farmers have developed onions that do not irritate the eyes, which are known as "Supasweet onions." These onions are grown in low-sulfur soils, and consequently the onions contain much less sulfur.

Why do our eyes sometimes water when we yawn?

When we yawn we pull unusual facial expressions. In doing so, we contract a number of facial muscles, and these contractions may put pressure on our tear glands, causing our eyes to water. These expressions may also put pressure on our salivary glands, and so also result in the mouth watering while we are yawning.

How do tears help to protect us from infection?

Almond-shaped structures called lachrymal glands are found above and slightly behind each of our eyes, and it is these glands that are responsible for producing our tears. When we blink, this stimulates the lachrymal glands, causing tears to be carried through the lachrymal ducts to the eye and eyelid. Tears keep our eyes moist and wash out dust and any other unwanted substances. Tears are sterile and contain an antibacterial enzyme called lysozyme, which is also found in saliva and human milk and which helps to protect us against infection occurring in our eyes.

We use the term "crocodile tears" to refer to a person who is insincere and whose tears are false. However, there is also a rare condition called "crocodile tear syndrome," which causes sufferers to cry in anticipation of food. This condition occurs most often after a person has suffered facial paralysis. What happens is that the nerve fibers attached to our salivary glands become damaged, and when they regrow they attach themselves to tear glands instead. As a result, instead of producing saliva in anticipation of food, the sufferer produces tears.

Why do our eyelids sometimes twitch?

Eyelids are two folds of skin that can be moved to cover or uncover our eyes. The upper and lower eyelids contain a row of eyelashes, which help to protect the eyes from dust and other substances.

Eyelid tremors, or myokymia, is a common condition in which a few of the muscle fibers of the upper and lower eyelids contract irregularly, resulting in twitching of the eyelid. This can be caused by stress, tiredness, or too much caffeine.

Why do we wake up with gunk in our eyes?

Day and night, various substances, including oil, sweat, and tears, make their way into our eyes. Tears contain salts, sugar, ammonia, urea (which is also found in urine), water, citric acid, and certain chemicals that kill bacteria. Blinking helps to wipe away these substances from the eyes during the day, but as we don't tend to blink during the night, these substances accumulate near the caruncle, the corner of the eye. This eye gunk can be solid or sticky depending on the amount of water inside it; the higher the water content, the stickier it will be.

The caruncle is a small, round, pinkish fold of tissue at the inner corner of the eye, which is believed by some experts to be the evolutionary relic of a third eyelid. Many nocturnal birds and some reptiles, including crocodiles, have a third eyelid, called a nictitating membrane, which moves horizontally over the eye. This special thin membrane helps protect animals' eyes from dirt and debris. Even when this membrane is closed, its transparency allows the animal to continue seeing.

Does the eye have a blind spot?

When we look at something, we are not aware that we have a "blind spot" in our vision. The blind spot is an area in our visual field in which we are not able to see objects. Here's a simple exercise to demonstrate the point. Hold up this book at arm's length and close your left eye. Then, focus on the dark circle in the box below and slowly bring the book toward your face. At some point, the X will disappear. This is because the X has moved into your blind spot. In practice, we are not affected by this blind spot, because we use both of our eyes in unison, ensuring a full range of vision.

On the surface of the retina, at the back of the eye, there are millions of color and light receptors, which are called cones and rods. These cones and rods convert light into millions of nerve messages. The nerve messages are then carried along the optic nerve to the brain, which turns these messages into a complete image.

The cones require a relatively high level of light to be stimulated, which means they function only in bright light, such as during the day. The cones see colors and so are responsible for our color vision. Rods, on the other hand, detect colors as shades of gray when it's dark. They don't need a lot of light to work, so you use them in dim light.

The part of the retina through which the optic nerve passes doesn't contain any cones or rods, so at this point the eye is blind to any images that fall upon it. This is the eye's blind spot.

Why do we produce earwax?

Our outer ear is made up of two sections: the pinna and the ear canal. The pinna collects sound and funnels it down the ear canal. The ear canal is curved (S-shaped) and about 1 inch (2.54 cm) long in adults. The ear canal contains glands that produce wax, which is also known as cerumen.

Earwax is made up of more than forty different substances, including wax, oil, and dead skin cells. Its main constituent is a substance called keratin, which is a protein found in the outer layers of the skin.

Earwax has various functions, which include trapping germs and debris to prevent them from reaching the eardrum and causing an infection. Second, it contains special antibacterial and antifungal chemicals to fight infection and so protects the ear from becoming infected. Third, it protects and moisturizes the skin of ear canal and so prevents the ears from becoming dry and itchy. The movements caused by chewing, swallowing, and talking help to move the earwax and any other debris, such as old skin cells, dust, or dirt, to the outside of the ear.

The exact composition of earwax varies from person to person, and it ranges in color from golden yellow to tan to dark brown or even black. People may have either dry or wet earwax, depending on their genes and other factors, such as the environment. Dry, crumbly earwax is more common in people of Asian and Native American descent, whereas moist and sticky earwax is more common in African Americans and Caucasians.

Is it possible for a spider to live inside your ear?

Although very uncommon, it is possible for a small spider or other small insect to take up residence in your ear.

In one example, a woman from Cardiff, Wales, went to see her doctor because she was suffering from itching and strange noises in her ear. When the doctor examined inside her ear, he was horrified to see something that, because of his magnifying equipment, resembled a tarantula. It was, in fact, a money spider, which had been there for around twelve hours. The doctor put ice water into her ear to force it out, and watched as the spider crawled out.

Another case was that of a Greek woman in Athens who complained to her doctor that she was suffering from headaches, as well as feeling a sharp pain in her ear when riding her motorbike. The doctor examined her ear and was surprised to find first a spider's web, and then a spider. The doctor said that the spider would have enjoyed its stay in the ear canal because the temperature was ideal for it.

Spiders are not the only creatures that might set up home in your ear. In 1997, a man went to his doctor as he was suffering from pain in his right ear. When the doctor examined the ear he found it was full of maggots. The man had fallen asleep on a beach a few days earlier, and it was thought that a fly must have laid some eggs in his ear.

What causes "red eye" in photographs?

"Red eye" can really spoil a photograph. With piercing red eyes, the cuddly family pooch can suddenly look like the hound from hell. Why does red eye happen? The reason is that when we take a photograph, if the subject is looking directly at the camera, the flash reflects off the retina, at the back of the eye. The retina contains many blood vessels that are red in color, and this is why our eyes appear red in photographs.

Many cameras have a red-eye-reduction feature, which causes the flash to go off twice, once right before the picture is taken and once moments later when the picture is being taken. The iris is the colored part of the eye, which controls the size of the pupil and the amount of light that is allowed to enter it. In bright light, the iris causes the pupil to become smaller, because the eye needs less light. The first flash causes

the subject's pupils to become smaller, so that there is a smaller opening when the picture is actually taken, which means the camera will not be able to pick up so much of the blood-filled retina.

Why are many babies born with blue eyes?

Many babies are born with blue eyes, which may later change their color. The colored part of a person's eyes is called the iris, and it is this part that controls the amount of light that is allowed to enter the eye. A baby's eyes are often blue, because they contain only a small amount of a pigment called melanin. Melanin is a brown pigment molecule—the more melanin that is produced in the eyes, the darker they will be. Melanin production generally increases during the first year of a baby's life, often leading to a deepening of the eye color. At around one year old, our eye color has usually settled on the color we will retain throughout our lives, but there are some cases in which eye color can change throughout adulthood.

Melanin colors our hair and skin as well as our eyes, and the amount and type of melanin we get is inherited from our parents. Irises that contain a large amount of melanin appear black or brown, whereas those that contain less melanin look gray, green, or light brown. If your eyes contain very small amounts of melanin, they will appear blue or light gray.

Why do albinos appear to have pink eyes?

Albinism is an inherited condition in which the pigment melanin is partially or completely absent. In some albino people, this may cause the eyes to appear pink or red, because light reflects off the red blood vessels in the back of the eyes. Non-albino people also have red blood vessels at the back of their eyes, but these blood vessels are usually not visible, because the iris is darkened with melanin.

Why did van Gogh cut off his ear?

Many readers will be familiar with the famous painting of Vincent van Gogh with a bandaged ear. Van Gogh painted this self-portrait in 1889, some time after cutting off his earlobe. But why did he do it?

Before he became an artist, van Gogh (1853–90) spent some time as a preacher. In 1888, he decided to leave Paris and move to Arles, in southern France. He invited the artist Paul Gauguin to stay with him and was overjoyed when Gauguin accepted. However, after a while the pair's relationship turned sour, and, while experiencing a psychotic episode, van Gogh threatened Gauguin with a razor blade. Later that day, van

Gogh is said to have felt such remorse for his behavior that he cut off the lower part of his own ear. He put it in an envelope and offered it to a local prostitute, who promptly fainted. Over the following months, van Gogh spent some time in hospitals and mental institutions, and in 1890, he shot himself fatally in the chest.

Do men really ogle more than women?

In any pub or club, you're like to see a stereotypical group of men drinking at the bar, ogling practically any woman who walks past—but is it only men who ogle? Surprising research shows that women will ogle an attractive member of the opposite sex even more than men will. Women have a wider range of peripheral vision, which means they are able to see a larger area than men. As a result, a woman can check out a man's body from head to toe without making it look obvious, so she won't get caught doing it. For a man to check out a woman, he has to look her up and down in a more noticeable way, which is why it is men rather than women who have the reputation for ogling.

CHAPTER EIGHT

Olfactory and Pulmonary Piffles

Is it possible to tell if someone is ill just by smelling him or her?

Doctors have known for hundreds of years that the smell of a person or the smell of his or her breath can help to give clues as to the condition from which he or she are suffering. Here are some examples:

- A person suffering with uncontrolled diabetes will give off a sweet, fruity smell. Diabetes is a condition in which the body is unable to use a hormone called insulin to process the sugar (glucose) in the blood properly. As a result, sufferers have too much sugar in their blood.

- A urine-like smell is produced when someone suffers with kidney failure.

- Measles smell like freshly plucked feathers.

- Arsenic poisoning smells like garlic.

How does a stuffy nose affect our sense of taste?

When we have a bad head cold, food seems flavorless. However, it is our nasal passages which are most affected by a cold, not our mouths, where the taste buds (receptors) are located. So why is our sense of taste diminished?

The reason is that our sense of taste is only partially based on the evidence of our taste buds. The tongue contains approximately 10,000 taste buds, which are microscopic structures shaped like tiny onions and which can detect chemicals in the foods we eat. The four main taste types are: sweet, salty, sour, and bitter. To produce our sense of taste our brains interpret data not only from the taste buds but also from our sense of smell. The nose has a large number of nerve cells, which contain receptors for many different chemicals. Foods such as strawberries and chocolate contain many chemicals that dissolve in the air and are detected by these nerve cells, which then send messages to the brain, where these smells are interpreted. So, not only do we taste the sweetness of the sugar in chocolate or strawberries, we also smell the odors associated with those foods. It's the combination of these smells and tastes together that gives us the flavor of "chocolate" or "strawberry."

When we have a cold our nasal passages may become swollen, and the nose may become blocked because of the extra mucus. Consequently, chemicals emanating from the foods we eat won't be able to reach the nerve cells in our nose. Because we are unable to smell these odors, our sense of taste is dramatically reduced.

To demonstrate this point, try this simple experiment. Close your eyes and hold your nose, and ask someone to give you a piece of either potato or apple, without telling you which it is. Believe it or not, you probably won't be able to taste the difference.

Why don't women have an Adam's apple?

The Adam's apple is hard lump found at the front of the throat. Both men and women have the same-sized Adam's apple until they reach puberty. However, when boys reach puberty they begin to produce much more of the male hormone testosterone. This hormone is carried around in the blood and causes the cartilage in the voice box to become bigger. The voice box (larynx) is an area in the throat that is connected to the top of the windpipe (trachea) and consists of a tube-like structure made of cartilage. Not only does the cartilage become bigger, but the voice box also tilts to a different angle in the neck, which makes the Adam's apple more prominent.

Why do we get hiccups?

Hiccups are sudden spasms of the diaphragm, the large, sheet-like muscle found between the chest and abdomen that helps to control our breathing. Hiccups can be caused in a number of ways, such as indigestion or eating too quickly. If you eat too fast, you may swallow air along with food, so the body expels this air in the form of hiccups. In other cases, hiccups can be caused by irritation of the nerves that run down from the brain to the diaphragm. The "hic" sound we produce is caused by the vocal cords snapping shut after each short, sudden gasp of breath.

People have different ways of getting rid of hiccups, such as holding their breath, drinking water from the opposite side of a glass, drinking vinegar, putting sugar on the tongue, biting on a lemon, or receiving a sudden fright. Some of these methods can occasionally help to get rid of the hiccups, but it's not clear why.

An American man called Charles Osbourne began hiccupping in 1922, and continued to hiccup for sixty-eight years, until 1990, when the hiccups mysteriously stopped. It didn't keep him from leading a full life—he married twice and had eight children.

Why do we sneeze?

Sneezing helps our body to get rid of potential irritants such as dust and pollen, and it also helps to clear our breathing passages. A sneeze happens when we feel a tickle behind the nostrils; then a nerve in the nose sends a message to the brain. The brain informs the muscles of the abdomen, chest, diaphragm (the large muscle under your lungs that assists breathing), vocal cords, throat, and even eyelids to work together in just the right order to get rid of the irritants, through the amazing mechanism of the sneeze. Our chest muscles squeeze the chest with enough force to shoot air up from the lungs and out through the nose at speeds of up to 100 miles per hour (160 kmph).

Sometimes bright light can make us sneeze—about one out of every three people sneezes when exposed to bright light. These people are called "photic" sneezers (photic means light), and this trait is hereditary.

Is eating boogers bad for you?

Doctors have a special name for everything. In the medical community, the technical name for using one's finger to pick boogers is rhinotillexis and the act of eating the boogers is called mucophagy.

There is an Austrian doctor who has gained notoriety by recommending that people, particularly children, should pick their noses and eat their boogers. Dr. Friedrich Bischinger, a lung specialist working in Innsbruck, believes that people who pick their noses with their fingers are healthier and happier as a result.

Dr. Bischinger has been quoted as saying, "With the finger you can get to places you just can't reach with a handkerchief, keeping your nose far cleaner. And eating the dry remains of what you pull out is a great way of strengthening the body's immune system." He added:

> Medically it makes great sense and is a perfectly natural thing to do. In terms of the immune system, the nose is a filter in which a great deal of bacteria are collected, and when this mixture arrives in the intestines it works just like a medicine. . . . Modern medicine is constantly trying to do the same thing through far more complicated methods; people who pick their nose and eat it get a natural boost to their immune system for free.

However, many doctors disagree, as they argue that the best way to prevent illness is to avoid germs.

Picking your nose ranks up there with spitting and burping for grossness and also involves health risks. Our fingers regularly touch items such as doorknobs and phones, which are crawling with bacteria as they are touched by numerous other hands. Touching a germ-covered doorknob and then picking your nose is an excellent way of transferring bacteria into your body. And since the inside of the nose is dark, warm, and moist, these microscopic germs will have the ideal conditions in which to live and multiply, which can lead to infection and illness.

Further, the nose's proximity to our brain creates another potential risk. If the skin inside our nose is broken when we pick it, the veins in that region are situated in such a way that sometimes an infection can travel straight to the base of the brain and interfere with the blood flow, causing a serious condition known as cavernous sinus thrombosis. This condition can also be caused by squeezing boils on or around the nose. Because of these risks, the triangular area of the face from the corners of the mouth to the bridge of the nose is often described as the "danger triangle of the face."

Why are boogers green?

The nose filters and warms the air you breathe before it reaches the lungs. It is lined with fine hairs and mucus (snot), which help to trap dust, pollen, and bacteria before they enter the lungs. Mucus consists mostly of water as well as salt, and chemicals that help it to stay sticky. Particles that get trapped in the mucus are moved out of the nose by small hairs called cilia, which move back and forth and can be seen only with a microscope. Cilia can also be found lining our air passages where they help to move mucus out of the lungs.

The body deals with millions of bacteria and viruses every day. When bacteria start infecting the nose and throat, white blood cells called neutrophils are produced in response. These cells contain proteins called human leukocyte antigens (HLA), and your immune system can recognize these proteins as "you." Any cells, such as bacteria, that do not have the right proteins are "not you" and are therefore attacked by the neutrophils. The neutrophils kill bacteria by eating them or by releasing toxic substances. Neutrophils are short-lived and often die during or after killing the bacteria.

When white blood cells meet germs such as bacteria, they make a large amount of an enzyme called myeloperoxidase, which is green because it contains a lot of iron. The coloration of snot therefore comes primarily from the iron. The breakdown of the bacteria is also thought to contribute to the color of snot.

Why do our voices sound funny when we inhale helium?

When you talk, your voice begins as a stream of air flowing from the lungs through the trachea (windpipe) until it reaches an area of the throat called the larynx (voice box). The larynx contains elastic vocal cords, and when the air passes between them they vibrate. Whether our voices are high-pitched or low-pitched depends on the length and thickness of these vocal cords. In men they tend to be thicker and larger, which accounts for the lower pitch of men's voices compared with women's.

One of the factors responsible for the voice's pitch is the speed of sound. In air, sound travels at around 1,090 feet per second (330 m/s). The speed of sound in helium is around 3,000 feet per second (900 m/s), which is almost three times faster. After helium is inhaled, it will travel from the lungs up to the vocal cords, and the sound waves will travel away from the vocal cords much faster in the helium, creating a voice that has a pitch nearly three times higher.

How does tobacco affect our lungs?

Tobacco contains sixty-three known cancer-causing substances, including arsenic, which is used in rat poison, and formaldehyde, which is used to preserve dead bodies. Smoking tobacco causes tar to enter our lungs; it is then absorbed by our lung cells, where it causes damage. The tar forms a brown sticky layer on the lining of our breathing passages, which can cause cancer.

The tar paralyzes or destroys the cilia in the breathing passages. Cilia are tiny moving hairs that sweep old mucus and dirt away from your lungs, like a broom. The mucus and dirt get transported to the throat, and if they are not spat out, they get swallowed and then destroyed by the acid in our stomach. Smoking just one cigarette can slow the sweeping actions of the cilia for almost an hour, and regular smoking virtually paralyzes them. If the cilia are paralyzed, the body will cough to try to clear the breathing passages. Over time, the cilia can become damaged, allowing tar to penetrate farther into the lungs, where it can do even more damage. This can lead to coughing, chest tightness, and shortness of breath. The good news is that cilia, which are paralyzed but not destroyed, can recover if the person stops smoking.

Although smoking is a dangerous habit, it did prove quite lucky for one American housewife. While out shopping, Steven Spielberg heard a lady ordering something. He was making the film *E.T.* at the time, and was having problems finding the right voice for E.T. himself. When Spielberg heard this lady's voice he knew she would be perfect for the role. The lady was an unknown American housewife called Pat Welsh, who had previously worked as a speech trainer, but excessive cigarette smoking had ruined her voice, leaving it croaky and gasping for breath—perfect for the voice of E.T.!

How do nicotine patches work?

A nicotine patch is like a big bandage, with an outer ring that sticks to the skin and an inner part that presses against the skin, slowly releasing nicotine into the body. Nicotine from cigarettes passes straight into the bloodstream through the linings of the lung, but nicotine in patches takes up to three hours to make its way through the layers of skin and into the bloodstream. The principle behind nicotine patches (and other forms of nicotine-replacement therapy) is that by providing nicotine in a form other than a cigarette, it's possible to reduce the symptoms of withdrawal while a person is trying to give up.

Have doctors ever prescribed cigarettes?

Nowadays we all know that smoking is bad for health, but up until the twentieth century many people believed that there were health benefits associated with smoking cigarettes.

Smoking was reputed to strengthen the stomach and act as a gentle laxative. It was also believed to prevent the membranes of the nose and throat from becoming inflamed. Doctors recommended smoking to calm the nerves and to help with insanity. Tobacco companies even advertised cigarettes by claiming they had health benefits such as "aiding indigestion," "restoring bodily energy" and, amazingly, "relieving fatigue to give athletes a competitive edge"! L&M cigarettes were advertized with the claim that their "alpha cellulose" filters were "just what the doctor ordered."

In China today, the government runs the tobacco companies, and in 2005, Chinese smokers paid $31 billion in tax. Until recently, the government promoted the health benefits of smoking, including a reduced risk of Parkinson's disease and the prevention of ulcers. They even claimed that smoking helped to prevent loneliness and depression. However, the government is now trying to reduce the number of smokers and has introduced antismoking campaigns. In China, around 360 million people smoke, and about 1.3 million die each year from smoking-related illnesses.

Why were children forced to smoke cigarettes at Eton College?

In 1665, the Great Plague struck Britain, and one of the recommended ways to ward off the disease was to smoke tobacco. Consequently, any boy at Eton College who was caught not smoking—and therefore not keeping the disease at bay—got whipped! A seventeenth-century English diarist called Hearne wrote:

> Even children were obliged to smoak. And I remember that I heard formerly Tom Rogers, who was a yeoman beadle, say that when he was a schoolboy at Eton that year when the Plague raged all the boys of that school were obliged to smoak in the school every morning, and that he was never whipped so much in his life as he was one morning for not smoaking.

What is "the bends"?

"The bends," also known as decompression sickness, is a very dangerous physical condition that can occur while a person is deep-sea diving. The bends can happen when a diver dives too deep for too long or comes up too fast. It is extremely dangerous, but thankfully it is also largely preventable. The way to overcome this problem is simply for the diver to resurface slowly.

The bends is caused by nitrogen bubbles being formed in the diver's bloodstream. The problem is that at high pressure, such as when a diver is deep in the ocean, the gases in the blood, such as nitrogen, begin to dissolve and turn into liquid. As the diver comes back up to the surface, the pressure on his body is relaxed, so the liquid turns back into bubbles of gas. If the diver comes up slowly, these bubbles appear in the lungs, and so can be breathed out. However, if the diver comes up too quickly, the bubbles will form not in the lungs but in the blood vessels. These bubbles can block the blood vessels, interfering with the flow of blood, and bubbles can also form between the bone joints. This causes intense pain, and sometimes even death. It is called "the bends" because one symptom is that the body becomes unable to bend at the affected joints.

How does castration affect the vocal cords?

In Italy between the 1600s and the early 1900s, one of the most popular types of male opera singer was the castrato. Castrati were men who had high voices like those of boy sopranos, but with the lung power of grown men. Unfortunately, the only way to achieve this kind of singing voice was for the castrato to undergo castration—in other words, to have his testicles removed—before he reached puberty.

During puberty, a boy's testicles produce increased amounts of the hormone testosterone, which causes the vocal cords, which are located in the throat, to become longer. Longer vocal cords cause the pitch of the voice to become lower, resulting in a deepening of the voice. Testosterone also stimulates the other physical changes associated with puberty, including enlargement of the penis and testes, the growth of facial and pubic hair, the increase in muscle mass and strength, and increase in height.

Most castrati were boys who had come from very poor families; their parents chose this unusual career for them in the hope that it would provide them with a better future. The boy would undergo a voice evaluation, and if it was decided that the lucky lad's voice had potential, his parents would arrange the castration. The doctor carrying out the castration would usually drug the child with opium or some other narcotic and place him into a very hot bath until he was nearly unconscious before finally chopping off the boy's testicles! Unfortunately, some boys died while undergoing this operation. Those who survived could never have children and were destined to have an unusually small penis.

Adult men who have been castrated have unusually long arms and legs, because castration affects the action of growth hormone and other growth factors. It also means these men develop little or no facial hair. They are also likely to develop excess fat on their hips, buttocks, and chest.

Why has no one found a cure for the common cold?

On average, adults catch two to three colds each year, whereas preschool children get more like six to twelve colds in the same period. The common cold is a viral infection, which can be caused by any of up to two hundred different viruses. When you catch a cold, this means the virus has managed to infiltrate your body's cells. In response, your immune system produces proteins called antibodies, which work to destroy the virus. After infection, your immune system keeps producing these antibodies for years to prevent you from ever catching that same cold virus again. However, since there are many other different cold viruses you will probably continue to catch colds, even though your body remembers how to fight all the ones you've had in the past.

There are two reasons why an effective cold vaccine hasn't

been developed. The first is that there are simply too many different kinds of cold virus. It would be unfeasible to produce a vaccine to deal with all of them.

The second reason is that, unlike bacteria, viruses actually invade the body's own cells. Because bacteria live outside the body's cells, bacterial infections can be cured fairly easily by drugs called antibiotics. However, colds are caused by viruses, not bacteria. Viruses enter and multiply inside the body's own cells, so an effective drug would have to kill the body's own cells in order to destroy the virus, and this would obviously be harmful. For this reason, antiviral drugs can only attack the virus immediately when it enters the body, before it enters the body's cells.

CHATTER

If you stuck your finger far enough up your nose, could you reach your brain?

No, because an average-shaped human finger would not be able to pass beyond the nasal passages. If a finger (and it would have to be unfeasibly small to do so) could make its way through the nasal passages, it would then reach the sinuses. Sinuses are air-filled spaces found behind the nose and cheeks and in the forehead. However, our imaginary finger would then find its route blocked by the cribriform plate, which makes up part of the ethmoid bone in the skull and forms the "ceiling" of your nose.

CHAPTER NINE

Amatory Arcana

How beautiful was Cleopatra?

In Hollywood films, Cleopatra VII (69–30 B.C.) is always depicted as being a legendary beauty, having been played by glamorous actresses such as Elizabeth Taylor. In reality, however, Cleopatra was apparently not much of a looker at all. Ten coins, which are all in good condition, have survived from the reign of Cleopatra. In one set of silver coins made in 32 B.C., the images of Cleopatra and her lover Mark Antony look remarkably similar. In another set, Cleopatra has an enormous neck and a long, hooked nose. It is believed that she inherited the hooked nose from her father, and this feature becomes more pronounced on coins issued later in her life.

The ancient historian Plutarch (c. A.D. 46–c. 120) was in an excellent position to report on Cleopatra's beauty, as his recent ancestors had served as her doctors and firsthand descriptions of the queen had been passed down through his family. According to Plutarch, Cleopatra was no striking beauty:

Her own beauty, so we are told, was not of that incomparable kind which instantly captivates the beholder. But the charm of her presence was irresistible, and there was an attractiveness in her person and talk, together with a peculiar force of character which pervaded her every word and action, and laid all who associated with her under its spell. It was a delight merely to hear the sound of her voice.

Even though Cleopatra may not have been beautiful, she was nonetheless able to seduce two of the world's most powerful men: Roman politician and general Julius Caesar, and Roman general Mark Antony, thereby saving her kingdom.

Why do testicles hang outside the body?

The testicles, which are also called the testes, are two oval glands located in a pouch of skin called the scrotum. The reason they hang outside the body is they need to be kept a couple of degrees cooler than normal body temperature, which is around 99°F (37°C), so that they can produce healthy fertile sperm. A higher temperature could lower sperm count.

The scrotum acts as a type of thermostat, which helps to keep the testes at just the right temperature for sperm production. In cold temperatures, the scrotum will draw the testes closer to the body for warmth. If the testes are too warm, the scrotum will loosen and allow them to drop farther away from the body, where it is cooler. If a man wears tight underwear, the scrotum is prevented from doing this because the underwear holds the testes close to the body. This is why wearing tight underwear can result in a lower sperm count.

Millions of sperm are produced by the testes every day, and they are so small that one drop of seminal fluid could contain more than 100 million of them.

Can testicles swell up to the size of watermelons?

The largest testicles on record are those of an African man whose scrotum weighed more than 150 pounds (68 kg) and measured almost 24 inches (60 cm) in diameter. He suffered from a disease called lymphatic filariasis, which is caused by parasitic, threadlike white worms, which are spread to humans by mosquito bites. These microscopic worms pass from the mosquito through the skin and travel to the lymph vessels. Lymph vessels are tubes that carry fluid called lymph throughout the body and make up part of the lymphatic system. The lymphatic system maintains the body's fluid balance and fights infections. The worms live in the lymph vessels, where they grow to adulthood. Adult worms live for around seven to ten years and grow to up to 4 inches (10 cm) long. When the adult worms mate they release millions of microscopic worms into the blood.

One symptom of this condition is called elephantiasis, which is the gross swelling of limbs and other body parts, such as the scrotum in males and breasts in females. This swelling is caused by a build-up of fluid. Male victims of elephantiasis may find that their testicles swell up to the size of watermelons, perhaps even bigger. The testicles and penis of one victim were so enlarged that he had to rest them in a wheelbarrow when he walked.

Worldwide, this disease affects about 120 million people, and is mostly seen in tropical Africa, Southeast Asia, and the Pacific islands. It is extremely rare in Western countries. This is because the disease cannot be caught instantly, but only after many thousands of bites from infective mosquitoes. Consequently, the process of catching this infection can take many months, or even years. Thankfully, there is a drug available that kills the worm larvae and thus reduces the swelling.

In 1881, a Scottish physician named Patrick Manson (1844–1922) discovered that the worms that cause elephantiasis are transmitted by mosquitoes. He had been working in China as a physician when he made the discovery, which was the first evidence that parasites can be spread by insects. It is said that his first encounter with elephantiasis was when he treated a Chinese peanut salesman whose scrotum was so large that he used it as a counter upon which to display his goods.

Why do men have nipples?

In the early stages of its development, the human embryo is female in its structure. As a result, men have some female features, such as nipples, mammary glands, and milk ducts.

Until the eighth week of gestation, the unborn baby is called an embryo. During this time, a certain gene in male embryos stimulates the production of the male hormone testosterone, which causes the embryo to develop masculine traits.

However, by this time the nipples have already formed. Testosterone causes the labia to fuse and eventually form a scrotum, gonads develop as testicles—and so a male is created.

Is it possible for men to breast-feed?

In theory, men could breast-feed, because they have nipples, small mammary glands, and milk ducts. They also produce oxytocin and prolactin, the hormones responsible for milk production in female breasts.

There have even been a number of reported cases of men breast-feeding their babies. It is claimed that some men were able to produce milk simply through extensive breast and nipple stimulation, either by using their hands or by letting a baby suckle on their nipples. One nursing father's milk was analyzed and was said to contain milk, sugar, protein, and other substances at levels similar to those of a mother's milk.

There have also been cases of men using hormone therapy to breast-feed. In Brooklyn, New York, a doctor reported that he helped a forty-year-old man to breast-feed his baby daughter. The unidentified man, a married transvestite, received the hormone estrogen to help develop his breasts before the birth of his baby. After the birth, the doctor administered oxytocin, which resulted in the release of prolactin. Prolactin stimulates the breasts to produce milk, and oxytocin helps the milk to flow from the breasts. The man was able to share breast-feeding duties with his wife and breast-fed for three months.

How does a penis become erect?

The penis becomes erect simply because it fills with blood. This happens when the arteries that supply blood to the penis become wider, so increasing blood flow into the penis. When a man becomes sexually aroused, his brain sends nerve signals instructing these arteries to relax their muscular walls. As a result, the arteries expand and consequently fill up with blood. This extra blood flow causes the penis to swell. Further, this swelling blocks off the veins that normally drain blood away from the penis. As a result, the blood pressure inside the penis rises dramatically, so the penis enlarges and stiffens. To make the penis go flaccid, the arteries that supply blood to the penis become narrower, causing a decrease in blood flow to the penis. As a result, the veins widen and begin to remove blood from the penis, and so the flow returns to normal.

How does Viagra work?

Viagra was originally intended as a treatment for high blood pressure and angina, until researchers found that there was an unexpected side effect—it caused an erection.

When a man is sexually aroused, nerve signals are sent from the brain to the penis. These nerve signals lead to the production of a chemical called cyclic guanosine monophosphate (cGMP), and it is this chemical that tells the muscles that line the penis's artery walls to relax and widen. Another chemical is responsible for breaking down cGMP, which will result in the loss of an erection.

The most common reason men, especially older men, suffer from erectile dysfunction is that, even though they might be aroused, there is not enough cGMP produced. As a result, the arteries in the penis fail to relax and widen, so the blood flow to the penis doesn't increase when the brain sends the signal. Viagra helps to prevent the breakdown of cGMP, which means it can build up in the penis, resulting in an increased blood flow. The more cGMP there is, the more blood flow there will be, meaning that an erection can be maintained for a longer time.

Which fish can get stuck inside a penis?

The candirú (*Vandellia cirrhosa*), which is also known as the Toothpick Fish (for reasons that will become clear), is a freshwater fish found in the Amazon river. This fearsome blood-sucking fish belongs to the catfish family and resembles an eel in shape. However, it is only 2½ inches (6 cm) long and transparent, which means it is very difficult to see in the water.

The candirú is strongly attracted to human urine and blood. If a person swims nude in candirú-infested water, the fish may swim into an orifice, such as the anus, vagina, or even the penis. If you urinate while in the water, this significantly increases the risk of a candirú's "homing in" on your urethra, the tube that links the bladder to the outside of the body. If the fish manages to swim into one of your orifices, it will erect its spine to lock itself in place and then begin feeding on your blood with its specialized teeth. Amazonian natives have developed various types of penis sheaths and G-strings made of palm fibers to protect themselves from the candirú.

According to reports, one twenty-three-year-old Brazilian man was attacked by the candirú while urinating in the Amazon river. It swam into his penis and got stuck, causing severe pain, fever, and bleeding from the penis. The dead fish was removed several days later, and the man suffered no long-term effects from the attack.

Can certain foods put you in the mood for love?

A well-fed stomach and an unclothed body breed
lust in a man.—*Chinese proverb*

Almost every culture on earth regards certain foods as having aphrodisiacal properties, meaning that they stimulate sexual desire. Before we understood the nutritional value of foods, people tended to assume that the best foods for love were those that resembled an erotic body part, such as the penis or the vagina. Rhubarb, bananas, leeks, and cucumbers have all been considered aphrodisiacs at some time, presumably because of their shape. Another well-known aphrodisiac is the oyster, which some say looks similar to a testicle, while others say it resembles the female genitalia. Oysters do appear to have some genuine aphrodisiacal properties, as they are rich in zinc, which is important for making sperm. The legendary Italian womanizer Giacomo Casanova (1725–98) reputedly ate seventy oysters a day.

The word "avocado" comes from the ancient Aztec word for "testicle." At harvest time, when the ripe avocado fruit were provocatively dangling from the trees, virgin Aztec girls were not allowed into the fields for fear the suggestive fruit would provoke uncontrollable sexual desire. However, very little scientific research has been carried out into aphrodisiacs, and so it is unclear how far any of these foods do actually stimulate erotic feelings. Perhaps simply the sight of rude-looking fruits and vegetables is enough in itself to stimulate feelings of lust?

Which parent's genes decide the sex of the child?

The human body is made up of around one hundred thousand billion cells (or, if you prefer, 100,000,000,000,000). Like the bricks of a house, cells are the building blocks of the body. Most cells, such as skin cells or muscle cells, contain forty-six chromosomes each. Chromosomes are strings of

DNA (deoxyribonucleic acid), which carry all the information needed to make an entire human being.

During conception, when the egg gets fertilized by a sperm, the sex of the child will be determined. Eggs always carry an X chromosome, but sperm may carry either an X or a Y chromosome. If the sperm carries an X chromosome the baby will be a girl. If the sperm carries a Y chromosome the baby will be a boy. Therefore, it is the sperm that decides the sex of a baby.

Can an ovarian cyst contain teeth?

The female reproductive system contains two ovaries, which are located in the pelvis, one on each side of the womb. Ovaries are almond-shaped reproductive glands in which the ova (eggs) are formed. The ovaries also produce hormones, such as estrogen.

An ovarian dermoid cyst is a bizarre tumor, usually benign, which develops in the ovary. It typically contains a variety of tissues, including hair, teeth, bone, skin, and even sweat glands. Some cysts have even been found to contain intestines and thyroid glands, and they often contain long strands of hair that bear no resemblance to the hair color of the woman from whom the cyst was removed. Cysts vary in size, but they average 2½ to 3 inches (6 to 8 cm) in diameter. However, they can become much larger and cause considerable pain. It's still unclear why cysts occur.

There have been recorded incidences of ovarian cysts growing to an enormous size, some weighing even more than a human baby. In 1991, a very large cyst was removed from the right ovary of a thirty-six-year-old American woman. The cyst weighed an astonishing 303 pounds (138 kg), and it was so big it had to be carried out on its own stretcher. The poor lady had been bed-ridden for two years prior to the surgery because of the sheer weight of the cyst, but she made a full recovery after it was re-moved. After the operation, she weighed just 210 pounds (95 kg).

What is the difference between identical and nonidentical twins?

Twins are a rare occurrence, which happens in just 1 percent of pregnancies. Around 30 percent of twins are identical, and 70 percent are nonidentical.

Identical twins originate from one egg which has been fertilized by a single sperm. The egg splits into two sometime during the first week after conception, and so develops into two babies who have exactly the same DNA, which means they are identical. However, identical twins do not have the same fingerprints. This is because at some point during the pregnancy, between weeks six and thirteen, the fetuses will touch the amniotic sac. The amniotic sac is a thin-walled bag that surrounds the fetuses during pregnancy. Touching this bag with their fingers affects the pattern of their fingerprints.

Twins who are not identical are called fraternal, or non-identical, twins. Fraternal twins do not come from the same egg; instead, two eggs are released and are fertilized by two different sperm. Both fertilized eggs travel to the uterus (womb), and both babies develop separately; each has its own umbilical cord and placenta. As each baby is a result of a separate egg and sperm, the babies will have different DNA and will therefore be nonidentical, so they will not look exactly alike. In fact, they will probably be no more alike than individual siblings born at different times.

Why do we have two kidneys?

Most people have two kidneys, but it is possible to lose a kidney—for example, through disease or injury. If a person does lose a kidney, the remaining healthy kidney can manage perfectly well on its own. The kidneys are bean-shaped and lie in the lower back. Their function is to filter our blood to remove waste products that aren't needed by the body, and also to help control the amount of water in the body.

When a person has only one kidney, this kidney manages to adjust so that it does as much work as two would normally. Each kidney contains about a million filtration units called nephrons—it is these that produce our urine. These nephrons can compensate for an absent kidney by increasing in size, so that they can do the work formerly done by two kidneys, therefore increasing the size of the remaining kidney. This happens with no adverse effects, even over many years. In fact, when people are born with only one kidney, it can grow to reach a size similar to the combined weight of two kidneys (about 1 lb or 0.5 kg). It isn't known exactly why we have two kidneys, but it is always good to have a spare!

Why is urine yellow?

Each kidney contains about a million nephrons, which are tiny filtration units in which urine is made. Our kidneys filter the blood so that substances not needed by the body can be expelled in the form of urine. The yellow color of urine is caused by a pigmented substance called bile, which is released by the gallbladder into the small intestine. Bile's function is to help break down fats that have been eaten. Urine varies in color from almost colorless to orange-colored, depending on how much water it contains. If we have drunk lots of water and are not sweating a great deal, our urine will usually be pale, almost colorless. However, if we are dehydrated through drinking too little water or through excessive sweating, our urine will be more concentrated and consequently it will be smellier and darker.

Urine can be odorless, or it may have a distinctive smell, which is usually described as earthy or slightly nutty. Urine is sterile when it leaves the bladder, unless the person urinating has an infection, but as soon as urine is exposed to air it starts to be broken down by bacteria that come from the outer part of the urinary tract and the skin. Bacteria cause the salts in our urine, especially urea, to be converted into a number of chemicals, including ammonia, which has a disgusting smell.

As well as our level of hydration, there are a number of other factors that can affect the color and smell of our urine:

- Reddish-brown urine may be caused by certain medications, by blackberries or beetroot in the diet, or by the presence of blood in the urine.

- Green, smelly urine may be caused by the consumption of asparagus.

- Dark orange, yellow, or brown urine may occur if someone is jaundiced, as he or she will have too much bile pigment in his or her blood, resulting in urine containing more of these pigments.

- Bright yellow urine may result from taking vitamin B supplements.

Some diseases can cause a change in the smell of our urine. For example, an infection with E. coli bacteria can cause a foul odor, while diabetes or starvation can cause a sweet, fruity smell.

Is it safe to drink your own urine?

Some people swear by a regular tipple of urine to keep them healthy. For thousands of years, in fact, people have been drinking urine for medicinal and religious reasons. During the 1600s, it was commonly believed that drinking one's own urine would help to treat depression, gout, toothaches, high fevers, and heartburn. In the 1700s, the standard French dictionary of drugs contained the following advice, "Fresh urine, two or three glasses drunk in the morning, fasting, is good against gout, hysterical vapors and obstructions."

Modern-day urine drinkers claim that drinking urine is effective against colds and flu, toothache, dry skin, baldness, and even cancer.

Drinking urine is unlikely to cause any harm to the body, because it consists of around 95 percent water, as well as other substances such as salts, vitamins, proteins, hormones, and antibodies (which help us to fight infection). However, the function of our kidneys is to remove toxins and waste from the blood. If we drink our urine, we reintroduce these waste products into the body. Nonetheless, the body can usually deal with them effectively.

Under normal circumstances urine is sterile (meaning that it's free from germs), unless a person is suffering from a urinary infection. In that case there will be bacteria present in the urine. Drinking infected urine could cause the infection to spread within the body.

A former Indian prime minister called Morarji Ranchodji Desai (1896–1995) began his day by drinking a glass of his own urine, as he felt it helped to boost his health. In 1995, on his ninety-ninth birthday, he attributed his longevity to the regular drinking of his own urine.

If you don't like the idea of drinking urine, perhaps you could take the advice given in the *Dictionaire Universelle des*

Drogues (Universal Dictionary of Drugs), which was written during the Renaissance by Nicholas Lemery. According to Lemery, "Human excrement is a digestive aid that helps dissolve, soften, and ease; it must be used in dried pulverized form and should be swallowed. A single dose should not exceed one dram [one sixteenth of an ounce]."

Why does asparagus cause smelly urine?

Within as little as twenty minutes of eating asparagus you may notice that your urine has an unpleasant smell. It's not clear why asparagus has this effect, but it may be caused by a substance called asparagusic acid, which protects the asparagus plant against parasites. There is a lot of asparagusic acid in young asparagus plants, and it is the young shoots that we like to eat. Our bodies convert the asparagusic acid into some very smelly chemicals, and these are passed out in our urine.

Around 50 percent of people who eat asparagus say they produce smelly urine after eating it, and 50 percent say they do not. Hereditary factors seem to play a part as to whether or not you will have smelly urine after eating asparagus. Perhaps only half of the people who eat asparagus have the gene for the enzyme that changes asparagusic acid into the smelly chemicals, or it could be that half the population is unable to absorb asparagusic acid from the gut. It has also been suggested that some people cannot detect the smell, so they too may be producing smelly urine, the only difference being that they are unaware of it.

Is it possible for a woman to become pregnant from having semen come into contact with her leg?

Sperm can live for up to seven days inside a woman's vagina. There, cervical mucus serves to nourish and protect the sperm. When sperm is outside the body, lacking this protection, it will die within four hours. Nonetheless, if semen is found on the inner thigh, very close to the vagina, there is a tiny, tiny possibility that a woman could become pregnant, because the sperm are capable of swimming in semen, so they could theoretically swim into the vagina.

Anyone who has touched freshly ejaculated semen should wash his or her hands before touching the vulva or vagina. Otherwise, sperm could be introduced into the vagina, possibly resulting in pregnancy.

Is it harmful to swallow sperm?

The average male ejaculation produces about a teaspoonful of semen, which contains up to 300 million sperm cells. Sperm cells make up only about 10 percent of semen; the remainder includes water and fructose, which is a type of sugar that helps to give the sperm the energy with which to swim. Semen also partially consists of a thin, milky alkaline fluid, which is produced by the prostate gland. This fluid helps the sperm to survive when exposed to the acidic lining of the vagina and uterus.

In general, semen is composed of proteins, fats, carbohydrates, and the mineral zinc, and if it is consumed the body will digest it and absorb it just like any other foods that contain protein, carbohydrates, fats, and minerals. Therefore, it isn't harmful to swallow semen. In fact, semen is also low in calories—a typical teaspoon of sperm will contain just five calories.

Why were clitoridectomies carried out in Victorian times?

In the prudish days of Queen Victoria, if a woman showed any interest in sex she risked being branded a nymphomaniac and mentally ill. When a woman was taken to an asylum a doctor would look at her clitoris, and if it was considered too big, she would be locked up and released only when the clitoris had become smaller.

There were various treatments for nymphomania, which included induced vomiting, bloodletting, straitjackets, separation from men, and leeches applied to the genitals. The most drastic measure was a clitoridectomy, the surgical removal of the clitoris.

A British gynecologist and surgeon called Isaac Baker Brown (1812–73) believed that any woman who masturbated should have her clitoris removed because, he believed, masturbation could lead to mental illness, epilepsy, and even death. In 1886, Brown's book *On the Curability of Certain Forms of Insanity, Epilepsy, Catalepsy, and Hysteria in Females*, in which he proposed that all of the conditions mentioned in the book's title could be cured by removal of the clitoris, was published.

Brown would carry out clitoridectomies by cutting off the clitoris, usually using scissors. He would then dress the wound and administer opium via the rectum. Brown claimed that his patients usually showed an immediate improvement. He declared that any woman undergoing this surgery would be transformed into a happy and healthy wife and mother within a month. Of course, if any of Brown's patients really did show an improvement in their nervous conditions, it was probably due only to the traumatic shock of the operation itself.

Unsurprisingly, the clitoridectomy proved ineffective at curing the disorders it was intended to treat. However, this did not prevent it from remaining a common treatment for a long time after Brown's death.

Can masturbation cause blindness?

Of course masturbation cannot cause blindness, but the Victorians certainly believed that it could. They also thought it could cause acne, epilepsy, mental illness, and even death. They went to great lengths to ensure that their children, particularly the boys, weren't self-indulging, and so a range of preventive measures were carried out, such as sewing up the foreskin (leaving a small hole for urination), tying the boy's wrists to the bedpost at night, or putting him in straitjacket pajamas. Some parents would even wrap a metal band attached by wires to an alarm around the penis at night. If the boy dared to have an erection, the alarm would inform his parents. Other ingenious antimasturbation devices include:

The Stephenson Spermatic Truss

This cunning contraption was patented in 1876. The penis was placed into a metal pouch, which was then stretched and pulled back between the legs and secured.

Four-pointed Penile Ring

This vicious-looking device speaks for itself—ouch! The device would be attached at night, to prevent the boy from having erections while sleeping. Thanks to the spikes located inside the ring, any hint of an erection would soon be stopped in its tracks. (Note the pretty bow!)

Bowen Device

This innovative device created in 1889 consisted of a cup placed over the head of the penis and attached to the boy's pubic hair by a pair of chains and clips. At the first sign of an erection, this horrifying contraption would yank out a mass of pubic hair—yikes!

Penis-cooling Device

This clever device, patented in 1893, ensured that if an erection did occur, cold water or air would be released through a tube. According to the inventor, "the cold water . . . cools the organ of generation, so that the erection subsides and no discharge occurs."

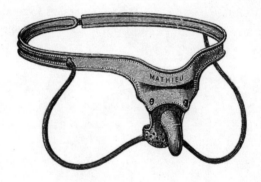

What did women use in the days before modern tampons and sanitary napkins?

Women in the past had a range of innovative ideas for dealing with their "time of the month." The ancient Egyptians are believed to have invented the first disposable tampon, which was made from softened papyrus (Egyptian writing paper made from the stems of plants).

In the fifth century B.C., a Greek doctor called Hippocrates recorded how ancient Greeks created tampons made of lint wrapped around pieces of wood. In Africa, women would roll grasses into absorbent bundles. In Rome, women used wool. Poor women in Victorian times used cotton rags, and later, Western women used washable pads of cotton or flannel, which were held in place by a contraption that resembled a diaper with a belt.

In America in the 1850s, a "menstrual protection device" was patented. It consisted of a belt fitted with steel springs to hold the pad in place; understandably, it took a while to gain popularity! In 1890, American women could buy the first disposable pad, which had loops at either end that attached to a special belt worn under the clothes. In the 1920s, the modern tampon was invented by an American doctor.

Even today, some women in poorer countries use natural sponges, rags, newspaper, or animal hides while having their period.

Is it possible to lose a tampon inside the body forever?

Tampons are absorbent plugs that are inserted into the vagina. Normally, the interior walls of the vagina stay in contact with each other unless something is placed between them, such as a tampon. At the end of the vagina is the cervix, which is the neck of the uterus (womb). It is a narrow passageway, which is too small for a tampon to pass through. For this reason, a tampon cannot get lost inside the body and can usually be removed quite easily. However, if a tampon is left inside the body too long it can cause an infection.

What were the earliest methods of contraception?

Around 1500 B.C., ancient Egyptian women used sponges, honey, or crocodile or elephant poo inserted into the vagina to help protect them from getting pregnant. The crocodile dung was often soaked in sour milk and this mixture would be stuffed into the vagina. The dung blocked the passage of sperm and also acted as a sponge, helping to soak up the sperm. Around 1000 B.C., Egyptian men would use a linen sheath as a means of contraception, as well as for protection against disease.

Between A.D. 100 and 200, when Roman soldiers were on long marches they would frequently have sex with local women and would use dried animal bladders as sheaths for protection.

During the 1700s, condoms made out of animal intestines (often sheeps' intestines) became available to the general public. These condoms would have a silk ribbon on the end with which to tie the condom to the penis. The condoms were quite expensive, so people would use them over and over again.

Casanova (1725–98) used sheeps' intestines to help protect himself and his lovers, and would blow them up like balloons to ensure there were no holes. He also recommended using half a lemon placed over the woman's cervix, because lemons are a mild spermicide. Although he was renowned for being a womanizer, he actually had sex with only 132 women, and when you consider that American basketball player Wilt Chamberlain (1936–99) claimed to have slept with 20,000 women, this number is relatively small. Casanova's lovers included servants, prostitutes, royals, and even nuns.

In the 1800s, the latex-based condom was made possible by the American Charles Goodyear (1800–60), after he invented the process of vulcanization. Vulcanization is a process that turns rubber into a strong elastic material. It meant that condoms could be mass-produced both quickly and cheaply.

Could a human being mate with an ape to produce a baby?

It's not clear whether or not human beings and apes could interbreed, but if they could, which is unlikely, the offspring would be infertile. To make a human being, forty-six chromosomes are required, whereas apes have forty-eight chromosomes. Horses and donkeys are similarly mismatched in terms of their number of chromosomes, but they can interbreed to produce mules. However, mules are, of course, infertile. As humans and apes also have a different number of chromosomes, a man-ape hybrid would also be infertile.

In 1977, researcher J. Michael Bedford discovered that human sperm could penetrate the outer membrane of a gibbon egg. Another scientist, a professor at the University of Hawaii, used salt-stored human eggs and found that hamster sperm could attach to and penetrate the shell surrounding the human egg. For ethical reasons the eggs were destroyed soon afterward.

How do pregnancy tests work?

One of the earliest written accounts of a urine-based pregnancy test can be found in an ancient Egyptian document dating from 1350 B.C. An ancient papyrus describes a test in which a woman who might be pregnant could sprinkle urine onto barley and emmer (wheat) seeds. If either grew, she would give birth. If the barley grew, she would have a male child, and if the emmer grew, it meant she would produce a female child. If neither grew, the woman was not pregnant.

This method was tested by scientists in 1963. They found that neither of the two seed types grew when watered with the urine of men or non-pregnant women. However, out of the forty pregnant women who were tested, twenty-eight of them produced growth in one or both of the seed types. The scientists concluded, therefore, that the seed test did indeed seem to be a good indicator of pregnancy. However, the reason why it works is unclear; one theory is that it is due to the variety of hormones produced during pregnancy, as some of these can induce early flowering in certain plants. However, there was no evidence for the claim that this test could indicate the gender of the child.

These days, pregnancy tests are extremely accurate. The most common pregnancy test involves a dipstick, which is a strip of absorbent material (a wick) on a plastic backing that is dipped into a sample of the woman's urine.

The way modern pregnancy tests work is that they look for a special hormone in the urine or blood that is present only when a woman is pregnant. This hormone is called human chorionic gonadotropin (hCG), which is the same hormone that gives the newly pregnant mother the feeling of morning sickness; hCG is produced in the body when a fertilized egg implants itself in the uterus. This usually happens about six days after conception. Most pregnancy tests use antibodies to detect the presence of hCG in urine. (Antibodies are proteins produced by the body's immune system in response to bacteria, viruses, or foreign substances.) Pregnancy tests contain antibodies impregnated in the wick, and if the urine, as it moves up the wick contains hCG, these hCG molecules will bind to the antibodies, producing a positive result. If there is no hCG present in the urine, there will be nothing for the antibodies to bind with, so the test will show a negative result. Pregnancy tests usually display a positive test either by producing a clear line or by a digital reading.

How young can a woman become pregnant?

Any girl who has a regular menstrual cycle can potentially become pregnant. The youngest known mother in the world was a girl aged just five years and seven months. Lina Medina lived in a remote village high up in the Peruvian Andes. When she was five years old her parents thought she was suffering from a massive abdominal tumor and took her to a nearby hospital. Doctors were amazed to find that the girl was pregnant, and in 1939, her child was born by cesarean section. The baby, a healthy boy named Gerardo, weighed 5 pounds, 14 ounces (2.6 kg).

Lina was thought to have suffered a hormonal condition related to the pituitary gland in the brain. She had been having regular periods since the age of two and a half, and by the time she was four years old she had developed breasts and pubic hair. It was uncertain who the father of the baby was, but suspicion fell on Lina's father. He was jailed for a while, but later released as there was a lack of evidence.

Gerardo was raised as Lina's brother, but at the age of forty he died from a disease that attacked his body's bone marrow. There wasn't thought to be a link between Gerardo's death and Lina's being such a young mother, or the fact that he may have been a product of incest.

The oldest mother on record is a Romanian woman called Adriana Iliescu, who, in 2005, gave birth at the ripe old age of sixty-six. After having IVF treatment she became pregnant with twins. At 33 weeks, one of the twin girls died, but Adriana gave birth to the other by cesarean section, a 3 pound, 3 ounce (1.5 kg) baby girl.

Can men grow breasts?

Gynecomastia is a condition in which men grow either one or two breasts. These breasts contain glandular breast tissue as well as fatty tissue, and studies show that the tissue found in gynecomastic male breasts is identical to that normally found only in women. There are also some men and boys who develop breasts that are made up of fat. This is a condition called "false gynecomastia" and can be helped by losing weight and exercising.

Both men and women produce estrogen in their bodies, but women generally produce higher levels. Gynecomastia is often caused by increased estrogen levels in a man's body, which leads to the sufferer's developing excessive growth of one breast or both. Some experts believe that this problem is caused by environmental pollution, through female hormones (estrogen) being leaked into the environment. This may be linked to the meat industry, which uses huge quantities of estrogens to make livestock grow bigger, ready for slaughter. Experts are also concerned that men are being exposed to estrogen through contaminated water supplies. One source of this contamination is the female oral contraceptive pill. Women who take the pill excrete in their urine small quantities of estrogens, which get into the sewage system and this can sometimes leak into the water supplies.

Gynecomastia sometimes occurs in teenage boys and is usually caused by an imbalance of hormones during puberty. It usually disappears without treatment within a couple of years. Certain medications can trigger breast growth in men, as can some rare genetic and hormonal diseases. It can also be caused by excessive alcohol consumption, using cannabis, or taking anabolic steroids. The condition can be treated using medication to reduce the extra breast tissue or, in rare cases, surgery.

Can a man become pregnant?

In 2003, a boy from Kazakhstan complained that he could feel something moving around inside him, progressively getting larger. Alamjan Nematilaev had had a distended stomach for many years, and when doctors eventually scanned his belly, they found what they thought to be a large cyst. When they operated, they found a round structure growing in his stomach. As they cut through the sac that surrounded the mass, they saw a head with dark hair, and a body complete with fingers and toes. This was not quite as shocking to them as it may sound to us, as cysts have been known to contain hair and teeth.

However, Alamjan did not have a cyst. Instead, he was diagnosed with a rare condition called "fetus in fetu," in which one healthy twin grows around the other. The fetus measured 8 inches (20 cm) and had remained alive by attaching itself to the blood vessels within Alamjan's stomach. The fetus received oxygen and food via the blood. Therefore, amazingly, the fetus was alive. However, it could only exist as a parasite, which meant it could remain alive only while it was inside Alamjan. As soon as the fetus was removed from Alamjan's body it would die, because it wouldn't be able to receive oxygen and nutrients via the blood.

In the past two hundred years, around only seventy cases of fetus in fetu have been identified, but it is believed this condition could be more common than previously thought, because people can have it without knowing it. Although these cases are extraordinary, they obviously do not constitute pregnancy, and, as yet, nobody is exactly clear why this strange phenomenon occurs.

Why were pregnant women advised to suckle a dog?

In 1825, America's first pediatric guide, the *Treatise on the Physical and Medical Treatment of Children,* was published. It was written by Dr. William Dewees (1768–1841), who advised expectant women in the eighth month of pregnancy to allow a "young but sufficiently strong puppy" to suckle at their breasts, to help toughen and accentuate the nipples, improve milk flow, and prevent inflammation. He claimed that nursing a puppy would "prepare the nipples for the future assaults of the child." He advised women whose nipples required some lengthening to take a long tobacco pipe and place its bowl over the aureola (the dark area around the nipple) and give it a good suck to help bring them out. As you might imagine, it is unlikely that any of these bizarre methods would actually have been of any benefit to the mother or baby.

In France and Germany in the 1500s, many children were suckled by goats, sheep, and other animals. If a mother or a wet nurse was not available to breast-feed a baby, the next best thing was for the baby to suckle at the teats of animals. In 1580, the French author Michel de Montaigne (1533–92) wrote, "It is common . . . to see women of the village, when they cannot feed the children at their breast, call the goats to their rescue."

Is it possible to be allergic to sex?

An allergic reaction is the response of the body's immune system to a substance that is usually harmless. People can be allergic to almost anything, and it is possible and relatively common to be allergic to latex in condoms, although an allergy to semen is extremely rare. When a man ejaculates, his semen contains millions of sperm plus proteins, sugar, vitamins, and minerals. If the woman's immune system recognizes one of these components as a substance to which she is allergic, her body will react by releasing chemicals called histamine. Histamine is present in cells throughout the body and is released during an allergic reaction. When the body has an allergic reaction to sperm, histamine is one of the substances responsible for the itching, burning, or swelling that can occur around the vaginal area, where the semen has been in contact with the skin. A small number of people may even suffer more severe reactions, including hives, generalized itching, and difficulty in breathing. It is even possible, although extraordinarily rare, for a woman to have such a severe allergic reaction that it can actually cause death. However, this allergy can be simply dealt with by using antihistamines, which are a group of medications that block the effects of histamine and prevent or reduce the symptoms of an allergic reaction.

In one recorded example, a twenty-five-year-old Romanian woman suffered an allergic reaction to the proteins in her husband's semen. She would choke and feel sick after having sex without a condom, so her doctors advised her to avoid sperm. However, the couple continued to have sex occasionally without a condom. On one of these occasions, the woman suffered a severe allergic reaction and was rushed to a hospital, where she died.

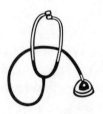